Sexual Selection: A Very Short Introduction

VERY SHORT INTRODUCTIONS are for anyone wanting a stimulating and accessible way into a new subject. They are written by experts, and have been translated into more than 45 different languages.

The series began in 1995, and now covers a wide variety of topics in every discipline. The VSI library currently contains over 550 volumes—a Very Short Introduction to everything from Psychology and Philosophy of Science to American History and Relativity—and continues to grow in every subject area.

Very Short Introductions available now:

Available soon:

For more information visit our website

www.oup.com/vsi/

Marlene Zuk and
Leigh W. Simmons

SEXUAL SELECTION

A Very Short Introduction

OXFORD
UNIVERSITY PRESS

Great Clarendon Street, Oxford, OX2 6DP,
United Kingdom

Oxford University Press is a department of the University of Oxford.
It furthers the University's objective of excellence in research, scholarship,
and education by publishing worldwide. Oxford is a registered trade mark of
Oxford University Press in the UK and in certain other countries

Published in the United States of America by Oxford University Press
198 Madison Avenue, New York, NY 10016, United States of America

British Library Cataloguing in Publication Data
Data available

Library of Congress Control Number: 2018940941

ISBN 978-0-19-877875-2

Printed in Great Britain by
Ashford Colour Press Ltd, Gosport, Hampshire

Contents

List of illustrations

Chapter 1
Darwin's other big idea

Anglerfish are some of the oddest-looking fish in the sea (Figure 1). They sport a wide mouth filled with needle-sharp teeth, above which dangles a slim, fleshy protuberance that gives the species its name and acts like a fishing lure to entice prey so that they can be gobbled up by these deep sea predators. When scientists first discovered anglerfish, they were puzzled that all of their specimens were female, although some had smaller apparently parasitic fish attached to their bodies. The parasites, a fraction of the size of the host fish, turned out to be not another species, but male anglerfish, living essentially as sacks of sperm waiting to spawn when the female is ready.

Other animals might not show quite such extreme sexual differences, but examples abound of males and females differing in appearance and behaviour. Male songbirds such as goldfinches are often more colourful than females, and are the sex that produces the musical song, while female spiders may be several times the size of males. Female baboons develop colourful rump swellings at the time when they are sexually receptive, and male bowerbirds construct elaborate arenas in which they enact vigorous performances of song and dance to attract potential mates.

The idea that males and females often look, sound, smell, and behave differently is uncontroversial. Where those differences came from,

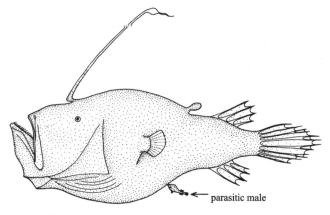

1. An anglerfish, with the tiny male almost invisible beneath the much larger female.

however, and what role they play in various species—including humans—is not. And just as Charles Darwin explained how natural selection could produce the variety of organisms on the planet, he also tackled the question of sexual differences in his 1871 book *The Descent of Man and Selection in Relation to Sex*. Many people assume that this book is about human evolution, but its primary focus is on his 'other' big idea in addition to natural selection: sexual selection. Natural selection, of course, is the differential reproduction of organisms based on their ability to survive, but sexual selection is differential reproduction based on sexual competition, whether between the members of one sex for access to the other, or by selection of particular mating partners.

Darwin was not completely alone in his investigations into sex, just as others had laid the groundwork for his ideas about the origins of biodiversity. A number of earlier scientists and scholars had pondered the question of why the sexes differ. Over two centuries before the publication of Darwin's major works, the English physician William Harvey, probably best known for his discovery of the circulation of blood, mused on the ability of a rooster, 'ornamented with his comb and beautiful feathers, by which he charms his mates to the rites of Venus'. Because the passions of sex, as well as war, were viewed as primitive and undesirable, some 18th-century philosophers attempted to determine the origin of such urges, and the famous writer and scholar Jean-Jacques Rousseau was determined to show that so-called savages in a state of nature were pure and good, while the baser desires were an invention of society. This separation of humans not only from other animals but from their ancestors may have contributed to the difficulty we have today in linking the behaviour of non-humans with humans.

Darwin himself was keen to understand the source of sexual differences in both. He carefully distinguished between traits used for survival, often the result of natural selection on the ability to find food or avoid predators, and those used in acquiring mates. He pointed out that while many animals exhibit extreme traits, in some cases these traits are found in both sexes and turn out to be beneficial in daily life, like the elongated curved bills of Hawaiian honeycreepers, which are used for probing flowers for nectar. But other traits, including those obviously needed for reproduction such as ovaries and testes, as well as the less apparently functional horns and antlers on sheep and deer, or the sweeping trains of male peacocks, only appear in one sex.

Differences in reproductive parts directly involved in sperm or egg production were relatively easy to explain, and Darwin called those primary sexual characteristics. They include the gonads and other aspects of internal and external genitalia. Although they are generally confined to one sex and exist at birth or hatching as they will throughout life, some animals may be hermaphrodites, in which the sex organs of both males and females occur in the same individual. In other animals, including some kinds of fishes and some invertebrates, an animal's sex—and sex organs—may change during its lifetime, so that, for example, a female blue-streaked cleaner wrasse will transform into a male if the male in her social group is removed. In clownfish, on the other hand, males may booome females if their mate dies or disappears. While some aspects of primary sexual characteristics also may be subject to elaboration, such as the structure of sperm or the morphology of a sexual organ, they have an obvious role in enabling gametes to come into contact and produce a zygote.

The other kinds of sexual differences, ranging from rooster combs to cricket chirps, were more difficult for Darwin to understand. In addition to not being needed to ensure that sperm meets egg, many of these characteristics are actually detrimental to survival, either because they require physiological energy to produce or because their conspicuousness makes their bearer vulnerable to predators. For example, in frogs and crickets, calling males expend up to twenty times more energy than they do when at rest. The long tails of boat-tailed grackles—North American birds related to orioles—make it difficult to fly in a high wind, while the bright orange colour patches of male guppies make them easy to spot by predatory fish in the streams where the guppies display. Darwin called these traits secondary sexual characters, and noted that in many cases they did not seem to have arisen through natural selection. How could the bearers of long and unwieldy tail feathers have been favoured by selection over their less elaborated counterparts?

Darwin said that sexual selection, which he distinguished at the start from natural selection, had led to their evolution. Like natural selection, sexual selection is a process that results from differential representation of genes in successive generations. In both processes, some individuals have characteristics better suited to the environment. With natural selection, those characteristics mean being able to elude predators, shelter from the elements, or find food. But merely surviving because of the ability to eat and not get eaten is not enough—if an animal lives but doesn't reproduce, it may as well not have bothered, at least from an evolutionary perspective. Survival of the fittest doesn't matter, unless the fittest also have offspring. Under sexual selection, then, the crucial traits that determine whether an individual reproduces depend on sexual competition, rather than survival ability. The secondary sexual characters, he proposed, could evolve in one of two ways. First, they could be useful to one sex, usually males, in fighting for access to members of the other sex. Hence, the antlers and horns on male ungulates, like bighorn sheep, or the aptly named rhinoceros beetles. These are weapons, and they are advantageous because better fighters get more mates and have more offspring, passing on their combative prowess to their sons.

The second way, however, was less intuitive. Darwin noted that female animals often pay attention to traits like loud songs and elaborate plumage during courtship, and he concluded that the traits evolved because the females preferred them. Peahens find males with long trains (the correct term for the fan-shaped set of tail feathers) attractive, just as we humans do, and such characteristics are ornamental. The sexual selection process, then, consisted of two components: male–male competition, which results in weapons, and female choice, which results in ornaments. When the sexes differ in their appearance, we call them sexually dimorphic. Species vary considerably in the degree of sexual dimorphism; in the anglerfish, of course, dimorphism is extreme, while in many seabirds, for example, the sexes are almost indistinguishable.

The battle over female choice

While competition among males for the right to mate with a female seemed reasonable enough to Darwin's Victorian contemporaries, virtually none of them could swallow the idea that females—of any species, but especially the so-called dumb animals—could possibly do anything as complex as discriminating between males with slightly different plumage colours. Many of those who found Darwin's ideas about natural selection unacceptable were even less enthused about sexual selection, and their objections focused primarily on female choice.

The difficulties that scientists of the day had with sexual selection mirror some of those that still dog the concept, including the degree to which the principles of mate choice apply to humans as well as non-human animals. Although the publication of *The Descent of Man* was not met with the same opposition as *On the Origin of Species* over a decade earlier, its explicit (for the time) discussions of sex and reproduction were greeted with disapproval. The book was thought to encourage immoral acts, and to rationalize what people viewed as reprehensible behaviour. Darwin referred to animals having 'wives' as well as 'concubines', and the moral implications for humans were quite explicit, at least to readers at the time. The 19th century saw many efforts at explaining the differences between human males and females, partly as fuel in the burgeoning women's suffrage movement, and Darwin's book fell squarely into the debate. From the very start, parallels between animals and humans were drawn, as people strove to find 'natural' differences between the sexes. Social context was virtually always present, and then just as now, people used animals as examples of behaviour they thought people should or should not emulate.

The naturalist Alfred Russel Wallace, who also independently arrived at some of the same conclusions about evolution and natural selection at nearly the same time as Darwin, was particularly vehement in his objections. He, and many others, simply found it absurd that females could make the sort of complex decision required by Darwin's theory; it would require the female to possess an aesthetic sense like that of humans, an idea they were unwilling to accept. After all, according to the thinking of his time, even among humans only those of the upper social classes could appreciate finer aspects of civilization such as art and music, so it seemed ridiculous to imagine that animals could do something many people—particularly non-Englishmen—could not.

Darwin tried to emphasize that choice could be defined differently in different kinds of animals, and that insects were probably the 'lowest' forms in which one could expect to see sexual selection, including female choice. At the same time, he was reluctant to depart from conventional ideas about courtship in human societies, and suggested that at least among less 'barbarous' groups, men took the initiative in choosing, which fitted in better with the concept of women as passive and coy. At the same time, Darwin proposed that choice could help distinguish modern humans from their supposedly more promiscuous and primitive ancestors. Wallace and other scientists at the time differed from Darwin in subscribing to the more generally held view of a sharp distinction between humans and other animals, in contrast to Darwin's view that continuity across living things existed, even if humans had unique characteristics. This lack of human exceptionalism, so to speak, was one of the reasons that Darwin's ideas were so revolutionary, and why people continue to object to evolution as an explanation for the origin and diversity of life.

Furthermore, Darwin was emphatic that his theories applied not only to differences in morphology and appearance, but to behaviour, another extension of thought that made some of his contemporaries uncomfortable. Some writers at the time wryly noted that Darwin actually gave more credit for independent thought to animals than to women, who were presumed to submit to their parents or other outside forces in choosing a husband. The mere notion that female animals, and by extension women, would respond sexually to male advances was also viewed with alarm, and Darwin had several exchanges with his publisher over terms such as 'allure' and 'excite' in reference to females. And a few feminists argued that if women were given more power to choose, society would benefit, because women were viewed as being altruistic, as opposed to the more selfish and egotistical male. Regardless, numerous aspects of the theory of sexual selection proved to be controversial.

The idea of female choice in the context of Western marriage also came up against another idea popular in the late 19th and early 20th centuries: eugenics, which promoted the selective breeding of humans to increase the prevalence of desirable characteristics like intelligence, and decrease the likelihood of what was often called 'feeble-mindedness'. Women were thought to shape the future of humanity by choosing eugenically sound husbands, which meant that beauty and other frivolous criteria were not seen as important. Eventually, the ideas of eugenics were used to justify the concept of a 'master race', familiar to us in the form of Nazism. Even after the Second World War, when Nazism was denounced, sexual selection may have been tarred with the same brush, and this guilt by association may have been why it was subsequently neglected by biologists.

In contrast to Darwin, Wallace proposed that the sexes differed not because of female preference for more ornamented males, but because the excess energy and vigour demonstrated in male activity had to be manifested somewhere, and the production of bright colours or loud songs seemed like a logical outlet. He suggested that the ancestral state for both sexes was conspicuousness, and that subsequent selection for females to be cryptic as they sat on the nest or cared for their young produced the dimorphism we often see.

Wallace and others also questioned the reason a female might have for choosing one male over another. If the only difference between them was the secondary sexual trait, why should the female bother? Wallace's view on male vigour leading to bright coloration or other ornaments was not widely held, and does not hold up to modern scrutiny: for one thing, what exactly does it mean to have an 'excess of vigour'? How could such a property, even if it could be defined, lead to a particular physiological manifestation such as a long tail or a loud song? Despite the absence of an alternative explanation for the evolution of elaborate sexual ornaments, puzzlement over how female choice could operate nonetheless led to near neglect of the entire idea of sexual selection for the next several decades.

Sexual selection and the Modern Synthesis

Although a statistician named Ronald A. Fisher developed mathematical models in his 1930 book *The Genetical Theory of Natural Selection* about how a seemingly detrimental ornament could evolve in only one sex, few biologists were concerned with sexual selection at the dawn of the 20th century. In the first decades of the century, several biologists began to combine the new findings in genetics with Darwin's ideas, to produce what is often called the Modern Synthesis. George Gaylord Simpson, Theodosius Dobzhansky, Robert Ledyard Stebbins, J. B. S. Haldane, and their contemporaries developed a theory about how fast evolution occurs, the relative importance of natural selection and random processes such as genetic drift, and how new species formed. Their work formed the foundation of contemporary evolutionary biology, and is still widely cited today.

For the most part, however, these major figures were simply uninterested in sexual selection. When they discussed extravagant traits of animals at all, these were suggested to have arisen to allow females to find a mate of the right species. Mating with another species is usually disadvantageous because the resulting hybrids are often less able to survive and reproduce than the parental strains. For example, a male horse can mate with a female donkey to produce what is called a hinny, while a female horse and a male donkey can mate and have what is called a mule. But mules and hinnies can't themselves make any babies, and in other species, even the first generation of such a cross between two species is unsuccessful. So species recognition is clearly important, and the early evolutionary biologists simply dismissed any other role for mate choice. Even Ernst Mayr, the distinguished scientist who developed ideas about speciation in the early 20th century, did not give much credence to female choice as Darwin conceived it, saying, 'There are many difficulties [in distinguishing between the different possible roles of display traits] even when these characters have only a single function. For example, it is now recognized that many phenomena previously thought to promote intraspecific sexual selection are actually specific recognition marks.'

In retrospect, it seems surprising that the extraordinary variety of secondary sexual characteristics were all deemed to serve merely to distinguish one species from another. Why some animal groups, such as birds of paradise, seemed to require elaborate feathers in a rainbow of colours combined with energetic courtship displays while others, such as gulls or some sparrows, could discern an appropriate mate based on a difference in leg colour or a streak of grey on the head, was never explained.

Researchers in animal behaviour at the time were similarly dismissive of the mere idea of female choice, presuming that choice was a rational process that then by necessity was confined to humans. Animals, they noted, were uncivilized, and hence the process simply could not occur in them. Furthermore, it appeared to the early zoologists that female animals were passive creatures that mated with whichever male could dominate them and fend off his rivals. And physiologists focused on the mechanisms leading to sexual differences, such as hormones, rather than on the evolutionary processes that would have selected them in the first place. The famous zoologist Julian Huxley made a detailed study of courtship in the great crested grebe, a water bird, and concluded that the complex displays by both sexes before pairing occurred were needed to overcome the female's natural reluctance to engage in copulation (Figure 2). It would appear that his own biases about the nature of males and females led him to make assumptions that were not necessarily warranted, as we will see in the remaining chapters. Interestingly, although he like other

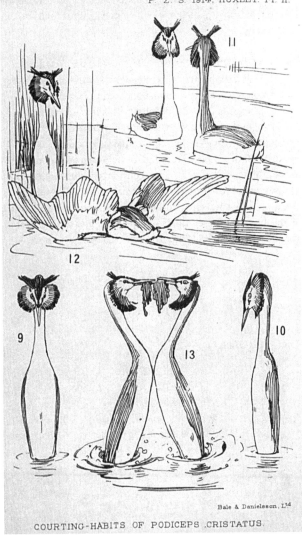

Bale & Danielsson, Lᵗᵈ

COURTING-HABITS OF PODICEPS CRISTATUS.

Darwin's other big idea

2. The courtship behaviour of great crested grebes is elaborate, and performed by both sexes.

scientists did not believe female animals could choose, Huxley did favour the idea that monogamy came naturally to animals other than humans.

Other efforts to examine mate choice in the laboratory or field were made during the time of the Modern Synthesis, often using the common representative animal for studies in genetics, the fruit fly *Drosophila*. When presented with a female, male *Drosophila* perform an elaborate courtship dance, sometimes accompanied by songs produced by rapid wing movements (Figure 3). Both sexes also respond to chemicals on the surface of the body. Males will often begin to court unsuitable targets, including immature females or other males, but will generally cease unless they perceive additional stimuli, such as the odour produced by a receptive female. It is possible that the flies were favourites for sexual selection research because it was more

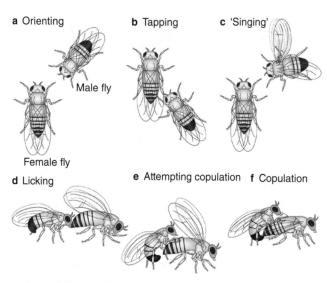

a Orienting **b** Tapping **c** 'Singing'

Male fly

Female fly

d Licking **e** Attempting copulation **f** Copulation

3. *Drosophila* courtship is remarkably elaborate, including a number of specialized movements and sounds.

difficult to anthropomorphize them than birds or other vertebrates, or perhaps they were simply more available for study.

Despite these early efforts, in 1948, Angus John Bateman published a still widely cited paper in which he pointed out that the evidence for sexual selection was 'mainly circumstantial' and attempted to determine why it was 'general law that the male is eager for any female, without discrimination, whereas the female chooses the male'. Bateman reported on experiments with *Drosophila* purporting to show that males continue to benefit with increasing numbers of matings, while female reproductive success plateaus after only a few mates. He concluded that because a male can fertilize many females, but females can lay only a limited number of eggs, male variance in reproductive success—the degree to which males differ in how many offspring they sire—is expected to be greater than female variance in reproductive success. This then suggested that sexual selection would act more strongly on males than on females.

Other mid-20th-century researchers on mate choice in fruit flies found some of the earliest evidence for genes influencing behaviour, another topic that continues to be hotly debated. They also described in detail the complex variations in mating behaviour that occurred under different circumstances. Claudine Petit, a French scientist who took up her studies after serving in the French Resistance in the Second World War, was convinced that sexual selection had been needlessly neglected, and performed numerous experiments on several different strains of the flies, demonstrating that sexual selection could be responsible for genetic variation in wild populations. Perhaps because her work was published in French, it was not widely known. Influential scientists of the time such as Theodosius Dobzhansky, however, were still unconvinced that animals, whether insects such as the fruit flies or vertebrates such as Huxley's grebes, actually demonstrated anything that could be called discrimination or preference.

The importance of individual preference

There matters stood, more or less, until roughly the 1960s, when ideas about animal behaviour and evolution began to shift away from a species-centred view and towards the recognition that individual differences were important. Biologists seemed to pay more attention to the ways that selection acted on individual variation, rather than on the similarities among members of a species and the overwhelming need to avoid hybrid mating. In addition, people began to question the notion of a passive, acquiescent female, and to wonder how exactly female choice could be enacted. In Chapter 2 we examine the variation in how males and females associate during the breeding season, ranging from brief couplings with multiple partners to lifelong monogamy. We also see how the discovery that females mate with many partners, even in supposedly monogamous species such as songbirds, was made possible by modern genetic techniques. Variation in mating systems holds considerable implications for the operation of sexual selection.

Chapter 3 will give a contemporary view of mate choice and its occurrence in a wide variety of animal species, along with some examples of what females are choosing and why parasites and disease may play a role in the evolution of extravagant secondary sexual characteristics. In Chapter 4, we will examine the notion, implicit in many of the original ideas about sexual selection, that males and females have natural 'roles', with characteristic behaviour associated with each sex. In Chapter 5 we see that sexual selection continues long after the physical act of mating is over, as sperm compete inside a female's reproductive tract and females bias the paternity of their young by selectively using sperm from particular males. Chapter 6 explores the ways in which benefits to males that accrue as part of mating may be detrimental to females, and vice versa, processes that seem to have literally shaped the evolution of genitalia in both sexes.

Finally, Chapter 7 considers the importance of sexual selection in determining how species diversify—that is, why some groups of animals may stay the same over evolutionary time while others become new species.

Sexual selection has to do with many of the ideas humans hold dear: fidelity, partnership, care of the young, the appearance and behaviour of the sexes, and whether the interests of males and females ever truly converge. Far from suggesting that a single answer to the question of male and female nature exists, however, we hope that understanding sexual selection will show our readers just how variable the natural world can be.

Chapter 2
Mating systems, or who goes with whom, and for how long

Although some animals shed sperm and eggs that meet at random, in most species, males and females have a characteristic association during the breeding season. These associations are called mating systems. Most commonly, mating systems are described as either *monogamous*, with one male and one female paired more or less exclusively during the mating season, or *polygamous*, in which more than one member of one sex is paired with a single member of the opposite sex. Polygamy in turn is divided into *polyandry*, where one female is associated with more than one male, and its converse, *polygyny*, with multiple females associated with a single male. Sexual selection theory suggests that males will benefit more by multiple mating, because their reproduction is limited by the number of females they inseminate, not by the offspring they can produce. Hence, we should expect polygyny to be more common than polyandry, and monogamy to be very rare indeed, given the difference in the way sexual selection acts on males and females.

This all sounds very neat and tidy, and it was the conventional wisdom for several decades, nearly up to the end of the 20th century. But cracks started to appear in the foundation of our thinking about mating systems a few decades ago, and we now realize that the situation is more complicated than we had thought. To understand why, it is helpful to look at a study that began with a question that had everything to do with controlling a pest, and nothing to do with sexual selection at all.

Blackbirds and marriage

What is the best way to keep birds from eating farmers' grain? Red-winged blackbirds are familiar harbingers of spring in many parts of North America. The species gets its name from the red shoulder patches on the otherwise glossy black males, and in marshy areas throughout the United States and Canada, the males defend territories and wait for females to arrive and make a nest on one or another male's territory. Up to eight female red-winged blackbirds can nest in the same male's territory, making them fit the definition of polygyny, and for many years, scientists studied the characteristics of the males that attracted the most females, assuming that many females on a territory meant higher reproductive success for the territory-holder.

In addition to their usefulness to students of animal behaviour, red-winged blackbirds are numerous enough in some places to be pests, because they eat grain from fields, sometimes in great quantities. Back in the early 1970s, wildlife scientists suggested controlling the birds' numbers without objectionable methods such as poisoning or shooting them, using a rather unusual technique: vasectomizing the male blackbirds. Sterilized individuals should behave normally, and the eggs would be laid, but they would not hatch.

A group of biologists accordingly performed the operation on a test population of red-winged blackbirds in Colorado. Much to their surprise and dismay, however, the majority of the clutches produced by the females from vasectomized males' territories were fertile. The paper published on this failed effort has a suspicious mention of the possibility of 'female promiscuity' being at fault, and concludes that vasectomy would actually not be an effective means of keeping blackbird numbers in check. They did not recognize, however, that their results foretold a revolution in thinking about mating systems.

In an unrelated effort, many years later, a group of Canadian scientists used the then-novel technique of DNA fingerprinting to perform paternity tests on the red-winged blackbirds in a marsh in Ontario. They discovered something startling, though the wildlife biologists in the earlier study might not have been so surprised: many if not most of the chicks in a given blackbird territory were fathered by a male other than the territory owner (Figure 4). Their work is now buttressed by numerous other studies showing that among many different kinds of supposedly monogamous birds, many nests contain chicks fathered by more than one male.

In this chapter, we first trace the way that animal mating systems have been explained historically, and then consider how more

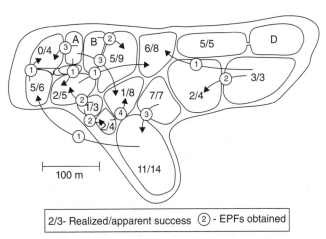

2/3- Realized/apparent success ② - EPFs obtained

4. A map of a Canadian marsh showing how the actual number of offspring sired by male red-winged blackbirds differs from the number of offspring in a male's territory. The arrows describe extra-pair fertilizations, with the origin of the arrow indicating the male that obtained the extra matings.

contemporary understanding of genetic and social relationships has reshaped our thinking.

Playing monopoly with mates

One of the first classifications of mating systems was made by biologists Steven Emlen and Lewis Oring, who said that a crucial variable is the degree to which access to mates can be monopolized by members of one sex. Such access can be direct—mates can be physically herded and then guarded—or indirect, by controlling access to resources such as food or nest sites. Individuals that need those resources then mate to gain access to them. Emlen and Oring suggested that this monopolization dictates the likelihood of acquiring one mate or many. For instance, if food is uniformly distributed in the environment, it is difficult to limit access to it, and hence polygyny is expected to be rare under such circumstances. But if it is patchy, then a male can sequester the resource and hence monopolize the females that go to that resource.

Emlen and Oring focused their attention on birds and mammals, and said that the red-winged blackbirds described above were an example of what they called resource defence polygyny. Although the birds do not feed in the marshes where the males hold territories, the nests are vulnerable to predators that eat the eggs and chicks, and better nest sites are those with lower risk of such predation. Males that defend good nesting areas within their territories are likely to attract more females. Females benefit from such an arrangement because they gain the resources provided by the male.

Emlen and Oring also pinpointed another essential contributor to mating systems, the ratio of males to females that are available for mating, called the Operational Sex Ratio or OSR. The Operational Sex Ratio is different from the simple proportion of males or females, because even if individuals of one sex or another are present, they may not be available as mates, perhaps because they are busy raising offspring or already have found a partner. A low OSR means that many females are available as potential mates, which should result in lower competition among males and a more polygynous mating system. Species that breed throughout the year will have a different OSR from those that breed in one concentrated bout, such as many frogs and toads.

Given this backdrop, we can see examples of the various mating systems in the birds and mammals that featured in Emlen and Oring's study. Polygamy of one kind or another, where at least one sex has multiple partners, is probably the most common mating system among vertebrates, with polygyny being more common than polyandry. Because of the competition among males for mates in polygynous systems, polygyny is often associated with very striking morphological or behavioural features in males that provide advantages in such competition. Some of the largest differences in size and appearance between the sexes are seen in polygynous species. For example, male elephant seals weigh more than three times as much as females, using their massive bulk to fend off rivals, and in birds of paradise, males are flamboyantly ornamented, with colourful feathers that are so elaborate they may impede the bird's ability to fly.

In some species, the male himself is not decorated, but he uses his environment to put on an extravagant show. Bowerbirds, a group of birds found in Australia and New Guinea, have some of the most impressive courtship displays in the animal kingdom, but they are not part of the male. Instead, male bowerbirds construct marvellous structures—sometimes several metres tall—out of branches and twigs, and then decorate them with colourful fruits, flowers, or stones. These bowers are used as staging areas for the complex dances and songs the male gives when a female alights nearby. Similarly, in some species of pufferfish, males build underwater 'sand castles', beautiful patterned arenas on the ocean floor that require a week or more of non-stop sculpting and shaping. Females lay their eggs in these arenas, and the eggs are then cared for by the male.

When the resources that females need can be monopolized by a male, as is the case for the red-winged blackbirds, the females have a dilemma: share a good territory with other females, or have a not-so-good territory to oneself? Depending on the form of the resource, it might be better to choose a less than ideal male or territory rather than divide the resource too finely, and indeed, it appears that a threshold exists, beyond which females divide themselves amongst the lesser-quality territories.

Even though sharing a high-quality territory with other females may make sense under some circumstances, males and females may maximize the number of offspring they have in different ways. In yellow-bellied marmots, large rodents found in mid-elevation mountains of North America, single females living by themselves after mating do not manage to raise as many offspring as those living with a male. If more than one female is associated with a male, each has a lower per-female reproductive success than a monogamous female. From the male point of view, however, having more than one female on his territory increases his reproductive success, up to three females, after which the number of offspring per male declines again. This means that in any one case, either the male or female does not reproduce at the optimal rate.

Extreme polygyny and leks

A somewhat unusual form of polygyny is seen in a variety of birds and mammals, which have group or communal display grounds called leks. The males gather on these areas, often using the same sites year after year, and the females come to a lek, survey the males, and mate with one of them. The lek is not a territory or a nest site, and the female gains nothing from the male besides the sperm to fertilize her eggs.

Leks are often spectacular to watch, as males with elaborate ornaments vie for the attention of the females. In parts of the North American west, sage grouse males gather each year in early spring on open areas in groups of 20–150 in the chilly pre-dawn to strut with their white chest feathers puffed out, inflating two yellow air sacs on their breast and then deflating them with a distinctive popping sound (Figure 5). Female sage grouse come to the lek individually, and appear to inspect various males. They may return to the lek on several days before selecting a male to mate with, after which they leave to lay eggs and rear the young on their own.

Other animals that lek (the term is used both as a noun and a verb) include hammer-headed bats, topi antelopes, and small tropical birds in the manakin family. Although the details of the displays vary, all leks have two things in common: communal displays, and the absence of any resource for females besides the opportunity to mate.

In most lekking species, just one or few males are responsible for virtually all of the matings. Hammer-headed bat males form leks in large trees along riverbanks in equatorial Africa; they display using loud honking calls and wing-flapping. In one study, 6 per cent of the males mated 79 per cent of the time. In white-bearded manakins, in one lek of ten males, a single male had 75 per cent of 438 copulations, and six of the males mated a total of only ten times.

These extraordinary behaviours and high rates of asymmetry in mating success prompt two questions. First, why do males cluster in these groups of displaying individuals? And second, why is mating success so highly skewed—if they gain nothing from the male besides sperm, why do females all seem to prefer the same mate?

To answer the first question, scientists have proposed three possibilities. According to the 'hotspot' hypothesis, males gather where females are likely to be found—hotspots—based on the

20.

1-3
Spruce Grouse
4-6
Ruffed Grouse
7
Sage Grouse
8-10
Prairie Chickens

5. Grouse males have a variety of ornaments, including the elaborate feathers shown here as well as a pair of air sacs hidden by the white breast feathers. The air sacs are used to make a 'popping' sound during male display.

females' usual routes to food or other resources. The 'hotshot' hypothesis suggests that it's not the places but the males that are the locus of attraction: most of the males surround a few very attractive individuals, and thereby profit from the females going to the hotshots. Finally, the female preference hypothesis holds that males cluster because females prefer sites with large groups of males. The females favour such groups either because it is safer, perhaps because they are shielded from predators, or because such clusters make it easier to compare potential mates.

To distinguish between the first two hypotheses, we can see what happens if a successful male at a lek is removed. If the hotspot hypothesis is correct, then removal of the male that had received most of the matings should result in the females mating with whichever of the remaining males moved into the site or sites where the previous winners had displayed. That would mean that the place, not the male, is key. Alternatively, if the hotshot hypothesis is correct, the entire cluster of second-rate males should leave, so they can find another hotshot, and females should not mate with those alternative males that do stay.

Tests like these in several lekking birds, including great snipe and black grouse, show that females tend to follow the male that had been attractive, or that the remaining males pick up and go when a successful male is removed. Both of these outcomes support the hotshot hypothesis. In other species, males do seem to track female movements and centre their lekking activities at those areas, supporting the hotspot hypothesis. The female preference hypothesis has received some support in yet other animals, including the sage grouse described above. As is often the case for complex behaviours, lekking probably has several advantages that contributed to its evolution, each of which is more or less important depending on the circumstances.

Given that males cluster, why do most or all the females mate with the same male? This question is at the heart of mate choice in general, but lekking species highlight the puzzle—what do females get out of mating with a male that is only giving sperm to fertilize their eggs? Possible solutions include the arbitrary and adaptive models that will be discussed in Chapter 3, with selection for parasite-free males being a possible criterion on leks. Indeed, male sage grouse with fewer parasites tend to get more matings. It has also been suggested that by mating with a dominant male on a lek, a female can avoid being harassed by subordinate males nearby, but little support exists for such a benefit.

Polyandry

Jacanas are medium-sized shorebirds that live in tropical parts of Africa, Australia, and Central and South America. Their elongated toes enable them to walk on lily pads and other aquatic vegetation, giving rise to local names like lily-trotters or even Jesus birds (Figure 6). At first glance, they seem to have an unremarkable mating system: larger, brightly coloured, and aggressive individuals defend groups of smaller drabber birds that care for the young. But in an unusual twist, the larger and brighter jacanas are the females, which fight with other females and lay eggs for each of the males that nests inside their territory. This mating system is called polyandry, the counterpart of the more familiar polygyny.

The relative rarity of polyandry, at least social polyandry, in which a female is associated for long periods of time with more than one male, is understandable. Presumably, in many if not most species, both females and males have little to gain by having multiple males associating with a single female, particularly when only one male fathers the offspring, since the female can't become more pregnant, so to speak, by mating many more times than is necessary to fertilize her eggs. Polyandry is also seen in several other species of shorebirds, in which females will mate with a male, lay a clutch of eggs, leave them with a male to incubate and rear the chicks, and then proceed to do the same with another

6. Jacana females fight with other females to defend territories where multiple males incubate the eggs and care for the chicks.

clutch of eggs and another male. In such situations, the female does not associate with all of her mates simultaneously, but more than one male is still mated to a single female.

Even fewer mammal species exhibit polyandry, though it occurs in gibbons and some types of tamarins, small tree-dwelling Neotropical primates. In such species, once the babies are born, the female hands them over to one of the males in the group, who carries them from place to place and protects them, essentially doing all of the care other than nursing the offspring. Polyandry even occurs in a handful of human societies, most notably in parts of India, Tibet, and Nepal, where women will sometimes take two brothers as husbands.

Why this mating system has evolved where it did is still unclear; some researchers speculate that at least in the shorebirds, polyandry may be an adaptation to very high rates of nest predation, such that females are under severe selection pressure to lay many clutches so that at least one survives. Of course, in both polyandry and polygyny, if parental care is needed for successful reproduction, a single parent must be able to provide most or all of that care. This limitation probably explains why polyandry is most prevalent in those birds with precocial young, including shorebirds, in which the chicks hatch able to move around and find their own food, rather than those with more helpless nestlings that require more intensive care.

One and one makes two

When a single male and female stay together for most or all of a breeding season, we call the mating system monogamy. Although it always means that the male and female are paired, the pairs might last a lifetime, as in tundra swans and a few other species, or just for a season or two, with partners switching in between. Often, the longer a pair remains together, the better their reproductive success.

Monogamy is often associated with parental care by both sexes, as in many songbirds, where the male and female both expend a great deal of effort bringing insects back to the nest for the chicks to eat. Although in general males benefit from multiple matings, as we have mentioned, monogamy can evolve when male help is essential for the young to survive. This may explain the high prevalence of monogamy in birds, with an estimated 90 per cent of bird species at least socially paired: their highly sophisticated flight capability means that a baby bird requires a great deal of care to become independent. Furthermore, once the egg is laid, a male bird can do everything a female can, by providing protection from predators and the elements and food for the chicks.

Of course, within monogamous species, males as well as females vary in how much care they give to the chicks; some males are particularly attentive while others seem to do only the minimum that is necessary. As we will see, the degree to which a male helps may be related to the prevalence of mating outside the pair.

Monogamy is relatively rare in mammals, though it does occur in a few species, including the gibbons, long-armed primates that live in the forests of Indonesia (Figure 7). Social monogamy may evolve if males cannot defend access to females because those females are widely dispersed in the environment, or, alternatively, may guard against infanticide by males. A handful of other animal groups in addition to mammals are also monogamous, including some you might not expect, such as a few species of cockroach and

7. Gibbons are among the few monogamous mammals, and the only monogamous ape other than humans.

some types of crocodile. Recent studies of monogamous rodents such as oldfield mice and prairie voles show that the tendency to pair bond is linked to genes that affect hormone levels in males and in turn their parenting behaviour. Injection of a hormone associated with nest building—an essential component of offspring care—can change the way that a male mouse behaves. Of course, this does not mean that hormones dictate the mating system, whether in mice or humans. But it suggests that selection acts on genes that control a variety of behaviours to produce the variation in mating systems that we see.

Rules, exceptions, and extra-pair mating

It's important to recognize that not all the individuals in a species or population will show the same mating system. In other words, even in a species we categorize as polygynous, some males may have multiple mates, others a single mate, and yet others none. Along similar lines, a given individual may not have the same number of partners over his or her lifetime, instead switching between multiple and single mates or remaining unmated.

To further add to the uncertainty, as we noted, the neat and tidy characterization of mating systems began to break down when DNA fingerprinting techniques began to be employed to determine paternity in wild animals. Birds were originally particularly useful in this regard because they have nucleated blood cells, so a tiny blood sample from a chick or adult can be used to get genetic material, although more recently DNA can be obtained easily from a variety of tissues. Birds were therefore among the first animals to illustrate the complexity of defining a mating system; in many species in which a single male and female appear to pair up during the breeding season, extra-pair mating occurs, and many clutches of eggs laid by a single female have multiple fathers. Birds vary in the extent to which they show extra-pair paternity, but at least some offspring sired by a male other than the apparent mate appear in 90 per cent of species.

What causes this variation among species is the subject of much research, with scientists examining the role of environmental variables such as seasonality and geography as well as the amount of genetic variation in the population as a whole. To reflect this realization, scientists now distinguish between the social mating system—what the social and behavioural associations between the sexes look like—and the genetic mating system, which determines who is passing on genes to the next generation.

While it is obvious how a male might benefit from fertilizing the eggs of a female in another male's territory, he also runs the risk of other males doing the same thing to him. How do males circumvent this problem? As we will discuss in Chapter 5, in many species mate guarding occurs after mating, whereby the male stays with the female and attempts to keep her from seeking additional mates. Mate guarding to prevent a female from mating with additional males is widespread, and occurs in animals with virtually all mating systems, not only the socially monogamous ones. Some butterflies guard their mates not by physically staying with the female, but by producing a substance along with the ejaculate that makes the female unattractive to other males, a sort of anti-aphrodisiac. In some animals, males will desert a female if she has been apart from him during the time before she lays eggs, perhaps because during that time she might have mated with another male.

From a female's perspective, engaging in extra-pair mating seems puzzling; if a single male can fertilize all her eggs, and if she risks having her mate desert during parental care, why would she seek out these extra-pair matings? Several different explanations have been proposed for the benefits to females of mating with multiple males or mating outside of a pair bond.

First, even though in theory a single male produces enough sperm to fertilize all of a female's eggs, that may not always be true, which means that mating with additional males can provide fertility insurance, reducing the risk that some eggs will remain unfertilized due to an infertile male. A related idea is that additional mates may ensure that at least one of the mates has genes that are genetically compatible with the female's eggs. This notion assumes that certain combinations of genes, not simply certain males, are particularly favourable, a subject we shall return to in Chapter 3. Second, if females choose their initial partner because he provides direct benefits such as parental care, they may be able to maximize their fitness by also mating with a male that is of high genetic quality but less likely to give such benefits. Some research suggests that females are picky about their extra-pair males, and that the characteristics that make a male suitable for a long-term bond may differ from those in a male that only mates with a female once.

In species with stable social groups, as in some primates and other mammals, females that mate with multiple males may obtain more direct benefits in the form of resources at the time of mating, or later, if males participate in parental care. This explanation is related to the idea of paternity confusion; if a female mates with several males in her social group, and males tend to invest in the offspring of females they have mated with, then no male can 'tell' which offspring he has fathered, and hence males will be more likely to help care for the young.

The ubiquity of multiple mating by females was recognized rather late in the history of sexual selection, although writers as early as Aristotle alluded to domestic hens mating with several roosters. Perhaps because many early researchers focused on birds and mammals, the idea that polyandry would be expected to be rare seemed logical to naturalists who only observed females with multiple mates in a few species such as the jacanas. Among insects and other invertebrates, however, multiple mating is the rule rather than the exception. And over the last decade or so, scientists have begun to reconsider the role of polyandry in sexual selection, going so far as to refer to a 'polyandry revolution' in the thinking of biologists.

Practical applications

It may seem as if the study of mating systems, while of interest from an evolutionary and behavioural perspective, would have no practical application. In fact, however, understanding a species' mating system can be critical in conserving its populations or understanding how likely it is to persist.

For example, hunters rarely kill animals without regard for their sex or age; often, larger males are preferred targets. It may seem as if this is a wise strategy, since the females that remain could presumably replace the members of the population that are lost, but the effect of male removal depends on the mating system. If a species is monogamous, and the young require care from both parents, for example, the loss of a male also may mean that the female cannot reproduce on her own, and the effect is the same as the removal of a whole family. Alternatively, if a male is removed from a species in which the males control groups of females, it may have a disproportionate effect on the population because multiple females are left without a male to father their young.

Using mathematical models to examine the effects of hunting in a wide range of mammal species in Tanzania, scientists found that killing only males would be much more likely to make a population decline than killing members of both sexes, particularly in monogamous species or those with male parental care. Using mating system information, the scientists concluded that for many species, including leopards, lions, warthogs, and several species of antelope, the quotas for hunting are higher than ideal for maintaining populations of the animals being hunted. Because the mating system is often linked to differences in mortality between the sexes, such as monogamous males living longer than highly competitive males in polygynous species, it is important to take these behavioural differences into account before constructing policies.

Similarly, the mating system influences the effective population size, a term biologists use to describe, not the actual numbers of individuals in a population, but the numbers that are actually breeding and contributing genes to subsequent generations. A species may seem to have a large population, but if only a fraction of the animals are breeding, then from a conservation perspective the population is much smaller, with potentially lower genetic variability, and potentially in greater danger of extinction than it might first appear. Again, polygynous mating systems can result in a much lower effective population size if a few males breed with many of the females, something that conservation managers would do well to take into account.

Chapter 3
Choosing from the field of competitors

Long-tailed widowbirds are small weaverbirds that can be found in the grasslands of central and southern Africa. Outside of the breeding season males and females are remarkably similar in appearance, having buff or tawny feathers with heavy streaking that helps them blend into their grassy surroundings. During the breeding season, however, males undergo a remarkable transformation (Figure 8). They shed their mottled brown feathers and replace them with feathers of jet black. Even more remarkably, they grow six to eight extremely exaggerated tail feathers that reach up to half a metre in length. Males conduct laboured flights across their territories with their tail feathers splayed and trailing beneath them, and are highly conspicuous to females and other observers in their environment over distances greater than a kilometre. Females will join males on their territories to build nests and raise young, typically fathered by the territory owner. In the early 1980s, Swedish behavioural ecologist Malte Andersson employed an experimental manipulation to provide definitive evidence that female choice is responsible for the evolution of sexual differences in plumage characteristics found in this widowbird.

Andersson counted the number of females nesting in each male's territory in an area of grassland on the Kinangop Plateau in Kenya. He then captured the territory owners and randomly

LONG-TAILED WHYDAH. ♂♂♀.
(Chera progne)

8. Male long-tailed widowbirds appear much like females outside the breeding season, but in the breeding season develop striking black plumage and elongated tail feathers and use these ornaments in displays to attract females.

assigned them to one of four treatment groups. In the first he cut the tail feathers to reduce their length by around 70 per cent, and in the second he glued the cut lengths of feathers to the tips of corresponding tail feathers of other males to experimentally lengthen them by around 50 per cent. The remaining two treatment groups served as controls, the first in which the tail feathers were cut and then reattached so that the tail remained the same length, and the second in which the bird was handled in the same manner as a manipulated bird but the tail feathers were left uncut. Andersson then released the birds back onto their territories and monitored the number of new nesting females arriving on each male's territory over the following month. He found that a disproportionate number of new nests were built on the territories of males with experimentally elongated tail feathers, compared with tail-shortened or control birds. This clever experiment provided direct and unequivocal evidence for Darwin's long-rejected suggestion that females choose among potential mates based on their secondary sexual traits, and shifted attention from the question of whether females exercised mate choice, to why they should exhibit the mating preferences they do. Thus, the last four decades have seen the development and empirical testing of a range of theoretical models for the evolution of female preferences and male ornaments, based on the notion that they can be either directly beneficial to females in terms of their own reproduction and survival, or indirectly beneficial through the action of genes inherited by offspring from their mother's chosen mate.

Choosing a good provider

Male stickleback fish build a nest in which females deposit their eggs, fertilized by the nest owner (Figure 9). Males provide exclusive parental care, fanning the eggs to promote gaseous exchange and protecting them from predation. During the breeding season males develop red coloration on the throat and chest and females prefer to spawn with more brightly coloured males. The intensity of male sexual coloration depends on the availability of carotenoids in their diet, the pigments that make carrots orange or tomatoes red. Males with access to greater dietary carotenoids not only develop redder throats and chests, but have greater longevity and provide better paternal care, increasing the chances that offspring survive to independence.

9. Male sticklebacks build nests (top left) in which females lay their eggs (top right). Males care for the eggs until offspring hatch. During the breeding season males develop nuptial coloration, including a striking red throat and chest, which females find attractive. Redder males are better fathers, have greater fertility, and produce offspring resistant to parasitic infection.

Female choice for male sexual coloration thereby provides direct fitness returns in terms of the quality of parental care provided to offspring. In general, theoretical models of preference evolution suggest that the direct benefits of parental quality can be a significant driver behind the evolution of female preferences, if male sexual traits convey reliable information to females on their ability to care for their young. As well as contributing to carotenoid-dependent nuptial coloration, dietary carotenoids can also serve as antioxidants, protecting males from the by-products of metabolism that induce damage to sperm cells and compromise their ability to fertilize eggs. In the case of sticklebacks, males with brighter nuptial coloration also have greater fertility, so that females choosing males with redder throats gain the additional benefit of ensuring their valuable eggs are fertilized.

Females often choose males to gain access to resources other than direct paternal care, such as nesting materials, food supplies, or other resources necessary for females to rear their offspring. For example, in many species of dragonflies and damselflies, females lay their eggs in areas of ponds and streams where the aquatic vegetation, water current, or water depth are optimal for egg development and hatching. Male dragonflies and damselflies will compete for access to these preferred egg laying sites, thereby monopolizing access to females that come to lay their eggs.

Similar mating behaviour has been recorded among birds, frogs, and fishes, with an extreme example in a bird called the orange-rumped honeyguide. Beeswax forms an essential part of the diet of these birds, and males establish year-round territories at exposed cliff faces in the montane forests along the Himalayas where bees nest. A small number of males control access to the limited supply of hives to which females come to feed. Courtship and mating occurs at the location of hives, and one study of this species documented a single male obtaining forty-six matings with eighteen different females. In such circumstances female choice of resources can drive intense male contest competition, indirectly favouring the evolution of armaments that contribute to a male's competitive ability. We discussed these so-called resource defence mating systems in Chapter 2. Female choice for such direct material benefits raises few conceptual problems, since it is obvious why females would prefer particular males. The most enduring question in research on female choice has been why, as in the case of Andersson's widowbirds, females prefer males with more exaggerated secondary sexual traits when those males offer nothing but their genetic material.

Sensory bias

Animals have finely tuned senses of smell, hearing, and vision, thanks to the intense natural selection acting on these sensory modalities in the context of avoiding predators and finding food. Trichromatic colour vision in primates, for example, is thought to have arisen in part because it allows primates living in a forest of green to find the ripe fruit or fresh vegetation that offer the greatest nutritional value. These sensory abilities can also, it turns out, be important in a completely different context, that of finding or selecting mates. If a trait that helps a female find food, for instance, makes her pay more attention to a particular colour or sound, then a male that exhibits such a colour or sound will be at an advantage over one that does not. This sensitivity to particular signals—sensory bias—means that sexual selection can favour males that exploit such biases. Water mites are sit-and-wait predators. They adopt what is referred to as a net-stance in which they grip aquatic vegetation with their hind limbs and raise their forelimbs into the water column where they can detect water-borne vibrations from their copepod prey. When a prey item swims past they will orient towards it and grasp it with their forelimbs. During mate searching, on encountering a female, a male will first tremble his forelimbs in front of her (Figure 10). The tremble frequency corresponds to the beat frequency of copepod swimming legs and attracts the female's attention such that she will orient to the source of trembling. Having attracted

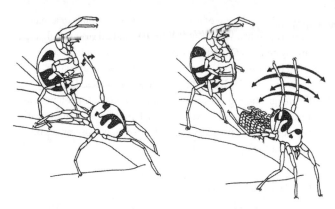

10. A male water mite encounters a foraging female in the net-stance posture (left). He trembles his forelegs to attract her attention before depositing spermatophores on the substrate in front of her. The male continues to tremble, mimicking the presence of a prey item to which the female responds, becoming inseminated as she does so (right).

the female's attention, the male deposits packages of sperm on the substrate before continuing his trembling. The female continues to orient to the source of trembling, walking over, and thereby picking up, the male's sperm packages as she does so. Starved females are more responsive to trembling, and are inseminated more frequently than sated females, which shows us that female foraging and male mating success are linked. Importantly, when researchers looked at the evolutionary origins of these traits, they found that net-stance evolved before trembling, consistent with the idea that male sexual displays can arise to exploit a pre-existing sensory bias in females. Similar conclusions have been drawn from studies of fish and frogs. In swordtail fishes, female preferences for an elongated lower lobe of the male caudal fin, the sword, pre-date in evolutionary history the origin of swords in males, while in túngara frogs female preferences for complex calls pre-date the evolutionary origins of complex elements in male breeding calls, suggesting that male sexual traits can arise to exploit sensory processes that are beneficial to females in a naturally selected context. However, male traits may sometimes result in responses by females at times or in places when mating is detrimental to them. We shall return to this problem again in Chapter 6 when we discuss the sexual conflict that can arise as a result of sexual selection.

Sexy sons

Sensory bias offers a mechanistic explanation for how female preferences for male secondary sexual ornaments might arise. A series of theoretical models have been proposed to explain how such preferences and traits might be maintained through the benefits they bestow on females and their offspring. The earliest theoretical model of preference evolution is attributed to the statistician Sir Ronald A. Fisher, who wrote several important works on evolutionary biology early in the 20th century. Fisher envisaged a two-step process. Imagine an ancestral pheasant in which females exhibit variation in their visual capacities that result in them encountering and mating with certain males, say those with longer tails, more readily than others, perhaps because of the sensory bias discussed previously. Suppose also that a mutation arises in the population that results in the development of a slightly longer than average tail in some males that confers a slight advantage to them, perhaps an increase in manoeuvrability during flight that enhances the ability to escape predators or reduces the energetic costs of flight. Females that have the pre-existing bias toward males with longer tails will produce sons with longer tails, capable of greater survival, and daughters with a preference for males with longer tails. The genes that encode long tails and preference for long tails will both increase in frequency in the population because of the selective advantage of the longer tails in flight. Once established, a second advantage pertains to sons with long tails, because males with the longest tails will have an increasingly strong mating advantage as the frequency of females with the preference increases in the population. Fisher argued that the linked inheritance of genes for preference and trait across generations would soon generate what he called a 'runaway process' which would see ever-increasing strength of female preference and exaggerated male trait such as that found in Indian peafowl. Indeed, the trait and preference for it could continue to coevolve to the extent that the trait becomes detrimental, now hindering flight or exposing males to predation, because of the benefits the male trait has in mate attraction. The currency of evolution is reproduction, not survival. The process will only be checked by natural selection when the survival costs of the trait outweigh the reproductive advantage that it bestows. While Fisher's runaway process was a verbal model, theoreticians have since developed formal mathematical models that confirm such a process can occur and there is experimental evidence from a number of animal species consistent with it.

Guppies are small live-bearing fishes that can be found in the freshwater streams of Trinidad. The males bear patches of bright orange, black, and iridescent blue while the female lacks any notable coloration. Among populations of guppies males vary considerably in the proportion of the body that is covered with orange patches, and females also vary in the strength of their preference for orange, such that populations in which males have more orange also have females with stronger preference for orange. Such correlated variation between male trait and female preference among populations is to be expected from the Fisher process, if the trait and preference coevolve in a runaway process until checked by natural selection against further exaggeration. It turns out that the populations of guppies with less orange and weaker female preferences contain a large predatory cichlid fish that feeds on adult guppies. The orange of adult males is also attractive to the predator, so that the evolution of both trait and preference are checked by natural selection against male conspicuousness in populations where predators occur, but are free to become more exaggerated in populations in which there are no predators of adult fish. Artificial selection for male orange in captive populations of guppies can cause an increase in the strength of female preferences for orange, demonstrating the genetic association between trait and preference upon which the Fisher process relies.

Good genes

Animals face a multitude of challenges in their lives. They must find food, disperse, avoid predators, have the energetic reserves to search, compete for, and display to prospective mates, and then they may have to raise the resulting offspring. The ornaments of sexual selection are typically greatly exaggerated structures that require the allocation of considerable nutritional reserves for their growth and maintenance. Thus, sexual ornaments typically exhibit condition dependence—that is, they are developed to an extent depending on the bearer's ability to allocate resources to their growth, rather than to other competing demands of life. In 1975 the Israeli evolutionary biologist Amotz Zahavi proposed that the ornaments of sexual selection could be viewed as handicaps, drawing resources away from traits necessary for an animal's general health and wellbeing. He argued that handicaps were revealing of a male's underlying genetic quality because males with good genes should be in a position to allocate relatively more resources to their ornaments than those with poor genes, allowing females to choose males offering genes for greater general health and wellbeing of offspring. When we talk of good or poor genes, we mean quantitative heritable variation, say in the ability to find resources in the environment, metabolize nutrients into energy, or raise an immune response to infectious disease, essentially all those traits that contribute to an individual's ability to survive, grow, and reproduce; in short, its fitness. While appealing, the idea of good genes sexual selection relies on the continued existence of genetic variation for fitness related traits. Ronald Fisher pointed out, however, that persistent selection on a trait would remove variation that can be inherited. Because fitness will be subject to strong directional selection there should be no genetic variation remaining for females to benefit from their choice of males.

The problem surrounding the maintenance of genetic variation in fitness is commonly referred to as the lek paradox. As we saw in Chapter 2, leks are communal display grounds to which females travel to choose among displaying males, but receive nothing from males besides the sperm to fertilize their eggs. They are common features of many animal mating systems. For example, in many species of birds, such as grouse, Indian peafowl, and cock-of-the-rock, the males gather together to display their sexual ornaments. The strength of sexual selection on leks is typically extreme, whereby a single male can often obtain almost all of the matings. If there is no variation in fitness among displaying males, why then should females have such consistent and strong preferences for one or a few of these males?

In 1982 William Hamilton and Marlene Zuk suggested that parasites might offer a resolution to the lek paradox. Hosts and their parasites are involved in continuous cycles of adaptation and counter-adaptation. Parasites are subject to natural selection to evade the host immune responses and promote their own growth and reproduction. At the same time, the host immune system is subject to natural selection to detect and destroy parasites that would otherwise compromise survival and reproduction. Host–parasite interactions are thus expected to generate cycles of adaptation between host resistance and parasite virulence, so that at any point in time there should be genetic variation in host resistance to disease. The theory predicts that parasites should have appreciable effects on the expression of male secondary sexual traits, females should choose males that are resistant to parasites, and, importantly, there should be heritable variation in resistance such that uninfected males sire offspring resistant to infection.

The stickleback fish we discussed earlier are infected by a variety of parasites that influence the intensity of red breeding coloration; uninfected males are able to develop the redder breeding coloration that females prefer. Moreover, males with a greater intensity of red coloration father offspring with greater resistance to infection by at least one of their tapeworm parasites, suggesting that male nuptial coloration does provide reliable information to underlying heritable resistance to disease. By exposing experimental populations of stickleback to parasites, researchers have found rapid changes in the frequencies of genes that regulate the immune system's response to pathogens. Moreover, resistance to parasitic infection arose after just two generations of exposure to a novel set of parasites. This latter finding demonstrates the potential for the cycles of adaptive coevolution thought to characterize host–parasite associations. In sticklebacks, then, females stand to gain direct and indirect benefits from their choice of redder males, in terms of direct paternal care and genes that protect offspring from parasitic infection.

It is now widely accepted that parasites can be important agents in sexual selection, although empirical evidence such as that found in sticklebacks remains largely incomplete for other animal species. Many studies have documented patterns of covariation within and among species between parasite loads and ornament expression, often with varying support for parasite-mediated female choice. Much less is known of the genetic basis of interactions between parasites and their hosts that would offer a resolution to the lek paradox.

The Hamilton–Zuk hypothesis is focused on a specific component of genetic variation—that of parasite resistance. A more general resolution to the lek paradox has been proposed to lie in mutation–selection balance. The so-called genic capture model for the maintenance of heritable variation relies on the fact that spontaneous mutations continually generate variation in male condition, defined as the ability of an organism to acquire and allocate resources to behavioural, physiological, and morphological traits that contribute to an animal's fitness. Thus, genes that encode resistance to parasites are just a small fraction of the genes that collectively affect a male's body condition. For example, genes that control other physiological processes also contribute to condition, such as the efficiency of metabolic pathways involved in the digestion of food, or the sequestering of digested products into fat reserves. Because condition relies on a very large proportion of the genome, it represents a large target for 'capturing' very small amounts of variation that are generated by random mutations across multiple genes (Figure 11). We saw above how the development of secondary sexual traits is often dependent on condition. By choosing males with more exaggerated condition-dependent ornaments, females may be able

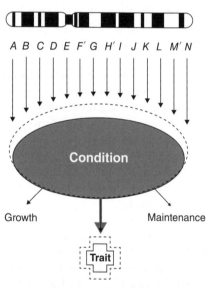

11. **A resolution to the lek paradox? Random mutations at multiple loci across the genome (e.g. at loci F, H, and M) all contribute to minor decreases in physiology or behaviour that reduce the amount of resources a male has available (condition) to allocate to growth, maintenance, and the development of secondary sexual traits. By choosing males with large traits females can avoid males carrying mutations. However, random mutations are reintroduced each generation, so that fitness variation is maintained by mutation–selection balance.**

to select genomes that are free of the mutations that would compromise the ability of their offspring to survive and reproduce.

Experimental studies have induced mutations in DNA to see whether spontaneous mutations can affect the expression of male sexual traits and whether such mutations can be subsequently removed by sexual selection. In guppies, for example, chemical mutagenesis of males led to delayed development and reduced courtship activity of their sons, but had no impact on the size or brightness of their orange patches. In dung beetles and bulb mites, mutations induced by ionizing radiation reduced the physical strength and survival of offspring respectively. In the beetles and mites, experimental populations exposed to sexual selection over several generations were returned to pre-ionization levels of performance while those in which sexual selection was prevented did not. So it would seem that sexual selection can be responsible for the removal of deleterious mutations, though the results obtained from similar studies with *Drosophila* fruit flies have been mixed.

Compatible genes

Offspring inherit two sets of genes from their parents: one from their mother and one from their father. Thus at each coding region of a chromosome, or locus, there are two copies of a gene (referred to as alleles) that together determine any given characteristic, the so-called phenotype, of the individual. The phenotype is thus determined by interactions between genes contributed by both parents. Some combinations of genes may be good, promoting offspring fitness, while others may be disastrous. For example, when both parents provide a gene containing a deleterious mutation, the resulting offspring will express the deleterious phenotype, resulting in poor development, infertility, or even death. Robert Trivers suggested that females should be expected to choose among potential mates based on their genetic compatibility, selecting males whose genotypes best complement their own in promoting offspring fitness. When females choose mates based on genetic compatibility, the lek paradox is not problematic because there is no directional selection for fitness. There is no single good genotype that females are seeking; instead, males that are preferred by one female will be rejected by others and vice versa.

There is good evidence that females in at least some species choose males based on genetic compatibility. For example, female vertebrates from fish and reptiles to humans have been found to prefer males with genes at the Major Histocompatibility Complex (a chromosomal region containing many genes important in resisting disease) that are different from their own. Such a preference provides offspring with a greater diversity of potential immune recognition molecules to orchestrate responses to a wider variety of parasites and pathogens. In at least some animals, females can use odour to distinguish MHC genotypes, because in addition to their role in the immune system, these genes are also associated with the production of odours released in the urine, sweat, or other body secretions.

Female choice for genetic compatibility and female choice for the most ornamented males need not be mutually exclusive phenomena. Indeed, work with house mice has revealed how females may weight the potential benefits from these good and compatible genes differently. Thus females choose males that scent-mark frequently, a trait that signals male social dominance and absolute genetic quality. They also choose males with dissimilar MHC genotypes based on odour cues contained in the same scent marks. But MHC dissimilarity seems to affect female choice only when the variation among males in their scent marking frequency is low. Female mice then prefer the strongest and most competitive males as mates, but when the field of competitors is narrowed, they will go for the male with the more compatible genes.

What about us?

Darwin devoted two chapters in his 1871 book to the role of sexual selection in the evolution of humans. He had 'little doubt that the greater size and strength of man, in comparison with women, together with his broad shoulders, [and] more developed muscles' were due to 'the strongest and boldest men having succeeded best...in securing wives, and thus having left a larger number of offspring'. He was also clear in his arguments that both female and male mate choice were important selection processes in the evolution of sexual dimorphism in humans. There is now evidence to support these views.

People consistently rate some faces as more attractive than others, and a variety of traits have been documented that contribute to face attractiveness. For example, both men and women find perfectly symmetrical faces more attractive than asymmetrical faces. We also find faces that conform to a population average configuration more attractive than those that deviate from the norm. Our faces exhibit obvious sexual dimorphisms. Men have a prominent brow, square jaw, and chin, all traits that develop at sexual maturity. Women find men with masculinized faces more attractive than men with more feminine faces, while the reverse is true for men. Our preferences for facial features are apparent across cultures and emerge very early in infant development, so they are unlikely to be driven by culture. Our preferences for sexual dimorphism also extend to body form. Women find men with broader shoulders and slimmer hips more attractive, while men find women with lower waist-to-hip ratios more attractive. And these preferences have an impact in our daily lives. Attractive individuals are perceived as being more trustworthy and are more likely to receive assistance when in stress. The biological basis for human preferences has been the subject of considerable research.

Sexual selection through mate choice appears to have been important in the evolution of human sexual dimorphism, and its signature has been detected in many populations. Men with more attractive, masculine features often report a greater number of sexual partners. While data are limited, some studies suggest that attractive men also leave more children, as Darwin suspected. Thus in a study of 1,244 women and 997 men born in Wisconsin between 1937 and 1940, men in the lower quartile for attractiveness had 13 per cent fewer offspring than their more attractive peers. Attractive women had 16 per cent more children than their less attractive peers. These differences in reproductive success were driven, in part, by a greater probability of marriage for attractive individuals. A similar study from Austria reported positive associations between women's facial attractiveness and number of children. Among the Hadza people of Tanzania, men with lower-pitched voices, which women find attractive, have been found to father a greater number of offspring. These studies show us that male and female preferences can impose selection on sexually dimorphic traits across human populations. Currently there is too little research to determine whether human mate preferences have evolved due to Fisherian runaway selection or because preferences target good genes. However, it is clear that in humans mate choice is a mutual process.

While females are typically the choosy sex, mutual mate choice is not uncommon. In the great crested grebes studied by Julian Huxley, for example, both males and females have the same exaggerated head ornamentation and male and female mirror precisely each other's behaviour during their complex courtship dance (Figure 2). The same is true of a number of bird species, including many seabirds such as albatross, penguins, and auklets. Cases of mutual mate choice have also been found in species of fish, frogs, and even insects. Mutual mate choice is expected to arise where both partners can benefit from their preferences. This is perhaps best understood in species where both sexes contribute to parental care, and both are limited in their reproductive success not by the number of mates they can acquire but by the number and quality of offspring they can raise together. Such mutual mate choice can lead to assortative mating for ornament expression, whereby the most ornamented pairs have the greater reproductive success, favouring the evolution of mutual ornamentation. Yet mutual mate choice need not necessarily lead to the sexes looking alike—humans are a case in point. In Chapter 4 we shall explore further the reasons behind deviations from the 'typical' sex roles in mate choice and in mating competition.

Chapter 4
Sex roles and stereotypes

Do we have 'standard' male and female roles? In humans, we often refer to sex roles as the attributes that are considered appropriate for each sex, and we know that components of these roles are influenced by the prevailing culture. For example, in centuries past, women were discouraged from seeking higher education in many parts of the world, because such learning was considered unfeminine and a violation of their natural role, but in at least some countries, this attitude has changed. Many references to modern human sex roles note the way that stereotypes about the sexes—women talk more than men, men have difficulty expressing their emotions—have the potential to limit our behaviour, and they also have social and political implications.

What about other species? Is there a typical male or female role in animals, one that could form the root of human sex differences? One version of sex roles holds that males are generally dominant and females submissive, stemming from the way that sexual selection favours different behaviours in each sex. This in turn could mean that sexual selection dictates particular behaviours in males and females. But in fact, sexual behaviour is extraordinarily varied in nature, though this does not mean that we cannot detect patterns of both morphology and behaviour in each sex.

Coy versus eager

As usual, we can start our consideration of sex roles with Darwin, who meticulously documented the behaviour and appearance of males and females of many different kinds of animals in his book *The Descent of Man*. He famously referred to females as 'coy' and males as 'eager' when it came to sex, and this categorization—possibly reflecting the social mores of the day as much as the biological facts—influenced subsequent biologists, as we noted in Chapter 1. In particular, Angus Bateman, whose work on fruit flies was mentioned in Chapter 1, suggested that females generally do not benefit by mating more than necessary to fertilize their eggs, while males continue to increase their fitness by seeking as many matings as possible. Bateman drew his conclusions both from Darwin's writings and his own experiments on fruit flies. He therefore considered sexual selection to act more strongly in males than females. Some scientists took this notion a step further and may still categorize the sets of behaviours that constitute sex roles to be either 'conventional' or 'reversed', with the former being characterized by choosy, even reluctant, females and showy competitive males, and the latter showing the opposite pattern.

Bateman influenced another important evolutionary biologist, Robert Trivers, who pointed out in an influential 1972 paper that we really do not have to discuss coyness or eagerness to explain the common patterns of male competitiveness and female choice. Instead, Trivers noted that females and males inherently differ because of how they put resources and effort into the next generation, which he termed parental investment. Females are limited by the number of offspring they can successfully produce and rear. Because they are the sex that supplies the nutrient-rich egg, and often the sex that cares for the young, they leave the most genes in the next generation by having the highest-quality young they can. Which male they mate with could be very important, because a mistake in the form of poor genes or no help with the young could mean that they have lost their whole breeding effort for an entire year.

Males, on the other hand, can leave the most genes in the next generation by fertilizing as many females as possible. Because each mating requires relatively little investment from him, a male who mates with many females sires many more young than a male mating with only one female. Hence, males are expected to compete among themselves for access to females, and females are expected to be choosy, and to mate with the best possible male they can. In some cases, depending on the kind of investment made, males might be choosy instead of females, and females may be competitive, as we will see in this chapter.

This, of course, is the same division of sexual selection that Darwin originally proposed, but Trivers gave it a new rationale. What he did, too, was bring female choice back to the forefront of sexual selection, and suggest a more modern underlying advantage to it. Furthermore, ideas about the evolution of animal behaviour had advanced enough that no one was worried anymore about an 'aesthetic sense' in animals; it didn't matter *how* females recognized particular males, just that if they did, the genes for the trait females were attracted to could become more prevalent in the population. Evolutionary biologist George Williams offered similar ideas in *Sex and Evolution*, published in 1975.

As for being coy, females do not necessarily refrain from mating or exhibit apparent reluctance during courtship; as discussed in Chapter 2 on mating systems, females from a wide range of animal groups mate many more times than is necessary to fertilize their eggs, and often mate with multiple partners even in supposedly monogamous species. Our closest living relatives, the chimpanzees and bonobos, are some of the most striking examples of multiple mating by females, and females even display their brightly coloured ornamental genitals to males, something

Darwin seems not to have realized. Furthermore, when scientists refer to 'sex roles' they may mean which sex is choosy about mates, which sex performs parental care, which sex takes the initiative in courtship or possesses the more elaborate secondary sexual characteristics, or some combination of these. Which roles are conventional and which reversed depends on the kind of animal you are discussing. In fishes, for example, males are as likely to care for offspring as females, so male parental care is not considered a departure from the norm. In certain shorebirds, on the other hand, females lay a clutch of eggs fertilized by a given male and then leave the eggs with their father, who incubates and protects them after hatching. Species that exhibit this latter behaviour might be said to exhibit sex role reversal, because the male usually does not play such a prominent role in parental care in birds.

With these caveats, however, we can still observe patterns across the animal kingdom that result from sexual selection operating differently on males and females. A survey of sixty-six animals, ranging from invertebrates to mammals, found that sexual selection did indeed appear to operate more intensely in males, and that this difference in intensity meant that males were more likely to be conspicuously ornamented compared with females. This pattern is in keeping with Darwin's and Bateman's ideas, although it does not need to indicate that females are coy in their behaviour.

When the shoe is on the other foot

One of the best ways to see whether generalizations about attributes expected in each sex hold up is to look, not at males or females per se, but at patterns of investment governed by sexual selection, as Trivers suggested. The key variable is how the members of a given sex leave the most genes in succeeding generations. That may occur when males are competitive and females choosy, which is the pattern that Bateman declared typical, or vice versa.

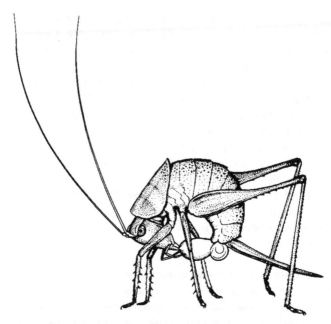

12. A female katydid with a nuptial gift, or spermatophylax, attached. The sword-like appendage is her ovipositor. Females eat these nutritious gifts that are quite costly for the males to produce.

For example, in most bush crickets (also called katydids), males supply a nutritious material, secreted from their own body's resources, along with the sperm when they mate (Figure 12). Females eat this material, called a nuptial gift, during or after mating, and the protein and other nutrients it contains help them produce more eggs, increase their lifespan, or both. The nuptial gifts are quite energetically expensive to produce, as they can weigh up to a third or even more of the male's body mass. To put this in perspective, imagine a human male having to manufacture something that weighed upwards of 15 kilos every time he had sex.

This high cost of the gift means that males cannot mate many times in succession, because they need time and energy to replenish their reserves. It also means that males could benefit not by the usual competition for access to females, but by being choosy about the females with which they mate. And indeed, in many types of bush crickets, males will preferentially mate with the fattest females, which are likely to contain more eggs and hence provide the male with more offspring. Females, in turn, may compete for access to the males. If the amount of food available in the environment changes, so may the behaviour of each sex; when food is plentiful, females do not compete as much for males and males can produce the nuptial gifts more frequently, while in times of scarcity, the opposite holds true, and the competition among females becomes more severe. This situation illustrates Trivers's model of sexual selection according to parental investment, since females in this case are investing less than males in any particular mating when food is scarce.

Similar behaviour can be seen in seahorses, shorebirds, and a host of other animals. In the seahorses, and their relatives the pipefish, males can become pregnant by fertilizing the eggs after the female has placed them in a specialized pouch on the male's body (Figure 13). The male develops a structure similar to the placenta in mammals to nourish the developing embryos. Some types of pipefish and seahorses are monogamous, with only one female providing the eggs, while in others, either or both of the sexes may mate with multiple partners. Females always give unfertilized eggs to the male, so that males do not care for offspring they have not sired. Female pipefish are often larger and more brightly coloured than males, and they court the males with elaborate displays that show off their ornaments.

A perhaps more obscure example of sex role reversal has been studied in small beetles that live inside seeds called honey locust beetles. In the wild, these beetles have actively courting females and choosy males that may reject females that try to mate with them. As in the bush crickets, honey locust beetle males provide

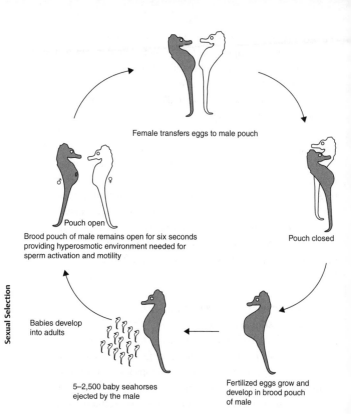

Female transfers eggs to male pouch

Pouch open

Brood pouch of male remains open for six seconds
providing hyperosmotic environment needed for
sperm activation and motility

Pouch closed

Babies develop
into adults

5–2,500 baby seahorses
ejected by the male

Fertilized eggs grow and
develop in brood pouch
of male

**13. Mating in seahorses. The male receives the eggs, and then fertilizes
them with his sperm before holding them in his body until they hatch.**

a nutritious nuptial gift along with the ejaculate. Females can
mate more often than males because males require time to
replenish those gifts. The researchers took laboratory populations
of the beetles and subjected them to 'experimental evolution', in
which some groups had five receptive females per receptive
male and others the converse, with five males and one female.
The latter thus experienced relaxed selection on female courtship
vigour, because males—and their gifts—were plentiful.

After nineteen generations in these treatments, male and female beetles from each type of group were observed after they were paired with a beetle of the opposite sex from a population that had not experienced either treatment. Females from the female-biased groups redoubled their efforts at courtship, making them more attractive to males. Somewhat surprisingly, males did not evolve many differences in their behaviour following the experimental evolution. And putting the beetles in low or plentiful food conditions was not as important as the scientists had thought it would be—apparently substances in the nuptial gift are manufactured by the male beetles themselves and just providing more food does not compensate for the resources acquired in mating.

Whether sex roles are conventional or reversed may depend on environmental circumstances, even within a species. In a small marine fish called the two-spotted goby, males care for the eggs, and more than one female may deposit her eggs with a given male. Such male parental care does not in itself constitute sex role reversal, as we have seen. Both sexes may reject mating attempts by the other, but early in the breeding season, male density is relatively high, and females are choosy; later in the year, males become more scarce, and several females may compete over the same male. Both sexes are ornamented and may perform courtship displays, and mutual mate choice occurs. The sex roles are thus flexible, and take different forms at different times of the year.

When both feet are in the same shoe

Hermaphrodites, in which male and female sex organs are found in a single individual, are an interesting test case for many of our ideas about sexual selection and sex roles. Hermaphrodites are seen in many plants, where pollen and ova (eggs) are produced on the same flower, but they also occur in a variety of animals, especially invertebrates. Hermaphrodites may be sequential, meaning that individuals change sex during their lifetime, as do some fish, molluscs, and sea stars, or they may be simultaneous, meaning that both types of sex organs appear in an individual at the same time, as in earthworms and some types of slugs.

Although it may seem as though hermaphrodites have the problem of mating solved, and that sexual competition would be unlikely in such animals, virtually all of the characteristics of sexual selection, ranging from competition over mates to the production of secondary sexual ornaments, have been observed in hermaphrodites. It is rare for hermaphrodites to simply join their own sperm and eggs to produce offspring; such selfing, as it is called, can cause harmful genetic mutations to become apparent, just as when close relatives mate and produce offspring. Instead, most hermaphrodites search for mates and then behave either as a sperm donor or an egg donor, with the other individual taking the complementary role.

Because sperm are generally cheaper to produce than eggs, it is sometimes argued that simultaneous hermaphrodites always benefit if they can take the male role in an encounter with another individual of the same species. And indeed, in some marine flatworms, individuals will actively remove the sperm they received from earlier matings once they have fertilized their eggs, and then will continue to mate as a sperm donor. Other marine worms engage in 'duels' with their penises as each partner attempts to inseminate the other while deflecting attempts at insemination by the other member of the pair.

A slightly more subtle way of influencing which individual functions as a male and which as a female is seen in many land snails, including the familiar garden snails. These hermaphroditic animals produce sharp calcified or chitinous 'love darts' as part of the mating process. Such structures are not sperm transfer organs; instead, they pierce a partner's skin and deliver

compounds that either inhibit the recipient from continuing
to mate or increase the fertilization rate of the sperm that
are introduced.

The advantages of being a male are not constant for hermaphrodites,
however, and other circumstances might favour either role.
For instance, if the likelihood of finding a mate is low, one might
do better to invest in ova, so that when a potential partner is
found, reproduction is maximized, since each egg will result in an
offspring, whereas each sperm is unlikely to find an egg to fertilize.
The moral of the story is that male and female functions cannot
be optimized at the same time, which means that hermaphrodites
must trade off the costs and benefits of egg and sperm production
against each other. Arguably, because in non-hermaphrodites
most of the same genes are recombined every generation to
produce both males and females, selection produces similar
compromises, since what favours a female—whether as a separate
individual or a component within a hermaphrodite—does not
necessarily favour a male. We discuss the conflict that arises when
something that benefits females (or being in a female role) is bad
for males and vice versa in Chapter 6.

And when many different shoes can end up at the same place

Even within a sex role, or a set of behaviours and morphologies
for a given sex, the strategies used to achieve mating may differ.
In many animals, some individuals of either sex may follow
so-called alternative reproductive tactics, behaviours that are
different ways of achieving reproductive success. For example, in
many fish, territorial males achieve fertilizations by chasing away
rivals and providing nest sites to females, but 'sneaker' males that
mimic females can also swim rapidly into the territory and deposit
sperm on eggs laid in the nest. These sneaker males are not as likely
to be detected by the territorial male, and they are able to achieve
some reproductive success without having to go through the effort

and risk of establishing a territory of their own. Such sneakers are seen in the two-spotted gobies mentioned above, and up to a third of the nests may contain eggs fertilized by such males. Other alternative tactics may involve being a 'satellite' male that intercepts females as they approach a territorial or signalling male. Both satellites and sneakers have been documented in a number of animal species, including insects and birds as well as fish.

Alternative reproductive behaviours come in two forms. First, individuals may follow the different behaviours—sneaker vs territorial male—at different times in their lives. For example, male elephant seals will guard groups of females at breeding sites for several months, chasing out all intruding males. They fight ferociously among themselves, and only one or a few individuals succeed in siring most of the pups in a colony. Sexual selection favours the largest males, leading to extreme sexual size dimorphism. But smaller, younger, subordinate males will sometimes sneak into one of the female groups and attempt to copulate, remaining inconspicuous and behaving rather like females. The sneaker males usually aren't successful because females often protest with special vocalizations when they are mounted, and this attracts the attention of the dominant bull elephant seal, who chases them away. Small males also attempt to copulate with females when they leave the colony, but these females usually aren't fertile. Nevertheless, younger males have no other options for gaining reproductive success, so they may follow the sneaker tactic and then switch to fighting for dominance with other males when they are larger.

In the elephant seals and other similar cases, the sneakers or other followers of the alternative behaviour are sometimes said to be 'making the best of a bad job'—they cannot gain any matings in the conventional way, by signalling or holding a territory, and so the alternative may yield at least some reproductive success. In such situations, one tactic might be best, but it is available to only a subset of individuals.

For example, in insects called scorpionflies, males can gain copulations in three ways. They can provide females with dead insects they capture—the better the gift, the longer and more successful the mating, as with the bush crickets already described. Obtaining prey is difficult, however, and males may also secrete saliva on leaves as a gift for the female. Finally, males may forgo the gifts and attempt to force copulations with females. In laboratory experiments that allowed males to use each of the tactics, it turned out that the best strategy (in terms of mating success) was to provide dead insects. However, this is only possible for large males capable of capturing prey. So the smaller males, which may be small because they were insufficiently nourished when developing, can only use the other two, less successful, tactics.

In the second form of alternative reproductive behaviour, individuals are genetically predetermined to follow a single mating behaviour, but several different mating strategies exist at the same time in a given population. One of the most remarkable examples of genetic control of alternative reproductive behaviours can be seen in the ruff, a medium-sized shorebird that breeds in marshy environments across Eurasia. Male ruffs occur in three distinct forms: a territorial aggressive one with a black or chestnut-coloured collar (or ruff) of feathers around its neck, giving the species its name; one with a white or mottled ruff that enters the territories of the aggressive males as a satellite and attempts to mate; and a female-mimicking one called a 'faeder' that was only discovered in 2006 (Figure 14). Just 1 per cent of males are faeders, and they have much larger testes than the territorial males, which, as we shall see in Chapter 5, probably helps them produce more competitive sperm that are favoured during post-mating sexual selection. The faeders also sneak onto territories just as females are crouching to solicit mating, and avoid detection via their female-like plumage.

The genetic basis for this unusually complex set of male morphs was published in 2016, when scientists used sophisticated

Independent   Faeder (female mimic)   Satellite

14. The genetically determined morphs of the ruff, *Philomachus pugnax*. Note the three different types of males in this species.

genomic techniques to discover that a single 'supergene' composed of 125 contiguous genes determines whether a male is a satellite or a faeder. The supergene appears to have arisen via an inversion, in which a stretch of DNA is flipped around, that occurred about 4 million years ago, a long time for such variation to have persisted. Two copies of the supergene are lethal, but the composition of the genes in individuals that have a single copy determines their morph. That first inversion seems to have differentiated between the territorial and other males. Then, a portion of the supergene seems to have inverted again some 500,000 years ago, yielding another variant. Males that have only the ancestral version of the stretch of DNA become the territorial morph, while males that carry one copy of the original inverted supergene become faeders, and those with the newest version are satellites. That a single stretch of DNA could influence male appearance, behaviour, and fertility is extraordinary, and appears to be due to the kinds of genes in the supergene, some of which affect levels of hormones such as testosterone.

A surprisingly similar example of genetically predetermined male strategies can be seen in an animal that is quite different from the ruffs, a small marine isopod, *Paracerceis sculpta*, that

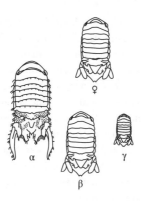

15. Male isopods come in three varieties, all pursuing different reproductive strategies.

lives within cavities of sponges in the intertidal zones along the Gulf of California in the Pacific Ocean (Figure 15). Like the ruffs, male *P. sculpta* come in three forms. Large alpha males are territorial. They occupy a sponge and allow females to enter, but attempt to exclude other males. Medium-sized males, the beta form, are female mimics, like the faeders. The beta males enter the sponges unchallenged by the alphas, and mate with the females that are already inside. Finally, the tiny gamma males slip undetected into the sponges, avoiding the other types of males that are present. All of these morphological and behavioural differences are controlled by a single gene with multiple alleles. When only one female is present inside a sponge, the alpha male sires nearly all of the young. But when two or three females are in a sponge, the other two forms can obtain some of the matings. Unlike elephant seals, who make the best of a bad job when they are young, the alternative morphs of these marine isopods are maintained in the population because, on average, they have equal reproductive success.

Other less dramatic but fixed differences among individuals occur in a number of insect species. In the solitary bee *Centris pallida*, adult females emerge from their clustered pupal burrows more or less synchronously and in a small area, which means that mating is extremely restricted in both space and time. The male bees compete fiercely for matings, but large males can always beat small males. Body size, however, is determined when the males become adult; small males do not grow into larger ones and body size is determined environmentally by the amount of resources provided by their mother. The males follow two different strategies. Large males are 'patrollers'; they fly rapidly over the female emergence sites and fight with other males for copulations with the females as they emerge. Small males are 'hoverers', and stay in a relatively small area where they attempt to mate with females once they are airborne.

Clearly, genes, the environment, or a combination of both can influence the evolution of alternative mating behaviours. When the forms are genetically determined, we usually assume that the various types must have equivalent reproductive success, since otherwise selection would have weeded out the inferior variants. In nature, however, the situation may well be more complicated, and sexual selection both drives the evolution of alternative mating strategies and is in turn influenced by them.

What about us?

Humans, of course, are also said to have sex roles, and while there are cultural influences on our attitudes and behaviour, as we noted at the outset of this chapter, many people have attempted to extend Bateman's principles to our own species. It turns out to be surprisingly difficult to obtain data about relevant items like the number of mates or children that people from different cultures are likely to have. And human populations present unique challenges for studying the existence or significance of sex roles from a biological perspective, if for no other reason than the near-impossibility of performing experimental manipulations. At the same time, being able to consult historical records or conduct interviews enables enquiries not possible for other species.

In keeping with Bateman's principle that males are expected to show more variation in reproductive success than do females, some studies have found that men from several different societies have a higher mean variation in the number of children they sire than females have in the number of children they bear. But societies around the world differ greatly in the degree to which they exhibit this tendency, or indeed whether they exhibit it at all. Certainly men in Western societies are more likely to remarry than are women, leading to a disparity in reproductive variability. A complication of this measure is that although one might assume that variance in reproductive success would have to be higher in polygynous societies, those in which a man may have more than one wife, the vast majority of men in such societies, over 95 per cent, are still monogamous, with a single mate. Even when men can marry more than one woman at once, women may have multiple partners over their lifespan.

Furthermore, as with the bush crickets and pipefish, when males contribute to offspring, having multiple mates does not always guarantee higher reproductive success. In many human societies, children are costly to both mothers and fathers, both literally and in the amount of time and energy involved in their upbringing. In others, the mother bears a disproportionate amount of the cost. Yet few studies of humans have detailed the relationship between partner number and number of children, a necessary prerequisite to testing Bateman's principle. This lack is particularly glaring when it comes to women, whether we examine pre-industrialized societies such as historical Finland or contemporary foraging societies.

Again as in non-human animals, factors such as the sex ratio may be important. A study of marriage patterns from 1910 in the United States revealed that in those states with a more male-biased sex ratio, the relationship between a man's socioeconomic status and his likelihood of being married was stronger, suggesting that women were choosier when more men were available. Other surveys of contemporary humans have found that when the sex ratio is female-biased, women may begin reproducing earlier, particularly in richer areas. Much work remains to be done on how human reproductive patterns may reflect differential selection on the sexes.

So do we do ourselves any favours by referring to 'conventional' sex roles, or 'role-reversal', or even to 'alternative reproductive behaviours', since the latter implies that a particular reproductive pattern is normal or standard? Certainly the terminology provides a useful shorthand for referring to consistent patterns that exist, albeit with exceptions, across an astonishingly broad range of animal species. Some scientists have advocated abandoning the concept of sex roles entirely, both because of the shifting meanings for different groups of animals and because of confusion between the culturally derived and biological notions. We suspect the terms are here to stay, but hope they do not keep people from acknowledging the enormous diversity of sexual behaviour across the animal kingdom.

Chapter 5
Sexual selection after mating

Darwin viewed sexual selection as a process that ended with mate acquisition, assuming that females are fundamentally monogamous, mating with just one male. This assumption, however, has turned out to be false. It was almost one hundred years after the publication of *The Descent of Man* before Geoff Parker's insight that sexual selection could continue after mating in the form of sperm competition. Parker pointed out that not only do female insects mate with several different males before laying eggs, they also store and nurture sperm from those males inside their reproductive tracts, in special storage organs called spermathecae. As a consequence, the sperm from several males must compete to fertilize the often limited number of eggs that the female lays. Birds were, until the molecular revolution, held as the paradigm of monogamous virtue. Now we know, thanks to DNA fingerprinting, that for the majority of species females will have chicks in their nests sired by one or more males that are not their social partner (see Chapter 2). It turns out that female birds can also store sperm from different males in tubules in the walls of their reproductive tracts and that sperm from different males must compete to fertilize available ova. Multiple mating by females has turned out to be ubiquitous across animal taxa. As a result, any trait that makes a male's sperm more likely to fertilize ova when in competition with other males will be favoured by selection because of the greater number of offspring he will leave in subsequent generations.

A male might increase his paternity share with a given female by reducing the chances that his partner mates with a rival male, or by influencing the fertilization process by placing his sperm closer to the site of fertilization, or tipping the fertilization odds in his favour by delivering greater numbers of sperm. Nevertheless, Parker recognized that females would not be passive vehicles within which males fight out their fertilization battles. Females may continue to choose among males after mating by differentially accepting, storing, or using sperm from more attractive males to fertilize their eggs. While sperm competition was envisaged by Parker as the post-mating equivalent of male mating competition, the post-mating equivalent of female choice has become known as cryptic female choice because it frequently involves processes occurring inside the female reproductive tract that are unobservable to us. In this chapter we will examine the far-reaching evolutionary consequences of sperm competition and cryptic female choice for the evolution of reproductive traits, from the gametes themselves to the adult organisms producing them.

Why are sperm so small and numerous?

A single human ejaculate contains, on average, around 200 million sperm. According to the *Guinness Book of Records*, the Emperor of Morocco, one Ismaïl the Bloodthirsty (1672–1727), fathered 888 children. It is estimated that Ismaïl would have needed to engage with one of his four wives and 500 concubines about once a day for 32 years, delivering in the region of 2.3 trillion sperm. That is 2.6 billion sperm per child. More usually, a monogamous couple not exercising any form of birth control might expect to have just 8–12 children during their reproductive lifespan, questioning the necessity for such high levels of sperm production. Another remarkable feature of sperm is their size. Eggs are typically among the largest cells in an organism while

sperm are typically among the smallest. Ostrich eggs can weigh
1.4 kg while their sperm are a microscopic 0.07 mm in length.

What evolutionary forces are responsible for such extraordinary
differences? The primitive animal condition is thought to have
been one in which both male and female produced gametes
of equal size, referred to as isogamy (anisogamy refers to the
opposite condition, when the sexes have gametes of different
sizes). Indeed, many marine invertebrates simply shed their
similar-sized gametes into the water column. Imagine an ancestral
species with isogamy. As in any biological system, we would
expect a normal distribution of gamete sizes (Figure 16).
The majority of individuals will produce average-sized gametes,
while some will produce smaller than average gametes and some

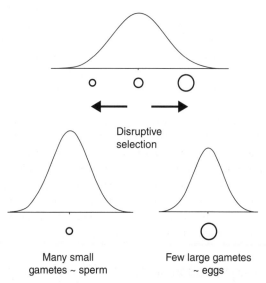

16. **Disruptive selection on gamete size can generate populations
of individuals producing either small or large gametes, and is thought
to have led to the evolution of sperm-producing males and
egg-producing females.**

larger than average. On a fixed energy budget, individuals producing small gametes will produce many more gametes than will those producing large gametes, because they require fewer resources to produce. Selection will favour small gamete producers because of their greater reproductive productivity. However, a zygote's survival depends on the amount of resources provided by the gametes for its development. We should therefore expect selection to favour individuals producing large gametes because their zygotes will be provided with the resources that allow them to reach independence. Whenever selection acts on the same trait, but in opposite directions, we term the net force disruptive selection, which can favour the evolution of alternative phenotypes, in this case small-gamete-producing males and large-gamete-producing females (Figure 16).

The stage is now set for sperm competition. Small gametes that fuse with other small gametes will produce zygotes with limited probability of survival. Any adaptation in small gamete producers that increases the probability that their gametes will fuse with the nutrient-rich large gametes will be favoured. However, because large-gamete producers are relatively less productive, there will not be enough large gametes available for all of the small gametes to obtain a fusion partner, leading to selection for adaptations that increase the probability that a small-gamete producer will fertilize the limited supply of large gametes. One obvious route to increased fertilization success for small-gamete producers is further increases in sperm number, with consequent reductions in sperm size. The more tickets an individual has in the lottery the more likely it is to win. Thus it is sperm competition that is thought to favour the evolution of numerous small sperm.

Sperm competition games

A considerable body of theoretical work shows how sperm competition should affect male expenditure on sperm production. In Parker's sperm competition games, as the theoretical models are called, a male can either expend its resources on gaining matings (for example on weapons and ornaments used to gain access to females) or on sperm production. Expenditure on sperm is assumed to increase a male's fitness in terms of fertilization success with a given female, but reduces the number of possible matings. Male fitness is therefore the product of the number of matings obtained and the number of offspring fathered from each mating. The most important point for our discussion here is that there are two assumptions behind sperm competition games. First, that males face a trade-off between expenditure on sperm and other reproductive activities, and second, that the fertilization success of a given male will depend on the number of sperm he has at the site of fertilization, relative to other males.

The predictions that arise from sperm competition games make intuitive sense. In Figure 17 the predicted expenditure on sperm (most often measured as testes size relative to body size) or conversely the predicted expenditure on weapons for mate

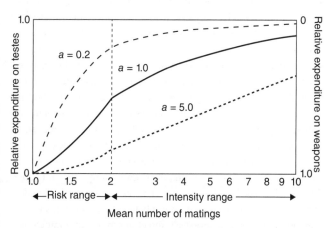

17. Theoretical models predict that increasing selection from sperm competition should favour increases in sperm production, and thus testes size, at the expense of decreased male expenditure on the weapons and ornaments used to secure matings.

acquisition are plotted against the level of sperm competition, being the mean number of males with which females mate in order to produce offspring. So, how much should a male be expected to invest, evolutionarily speaking, in weaponry or sperm production, given a certain probability that his sperm will be competing with that of other males? Note that there is both a *risk* and *intensity* range of sperm competition. In the risk range selection from sperm competition is low, with variation in the probability that a female will mate with just one other male. In the intensity range of the parameter space females always mate with two or more males. The models predict that as the risk and intensity of sperm competition increase, males should invest more in sperm competition and less on weaponry. But as the payoff from fighting for females increases (values of a), males should invest more in weapons.

Sperm competition affects testes evolution

Evidence to support sperm competition theory comes from many sources, including the very simple measure of how large a species' testes are relative to its body size. It was first noticed in primates that the mating system (Chapter 2) of a species can be a strong predictor of testes size and thus sperm production. A priori we expect larger species to have larger testes. However, variation within this expected relationship is best explained by the mating system (Figure 18). For example, gorillas form multi-female breeding groups with extreme male dominance. The silverback male (so named because of his age and hence greying fur) controls access to females in the group, and females rarely mate with a male other than the silverback. Gorillas have smaller testes than we would expect given their body size. At the other end of the spectrum, the pygmy chimpanzee or bonobo forms multi-male breeding groups. Females copulate with all males in the group. Indeed, copulation is a behavioural mechanism in bonobos for defusing aggressive interactions and maintaining social cohesion within the group. The consequence of course is that sperm from many different males are in the female reproductive tract at the time of ovulation, leading to intense sperm competition. As you

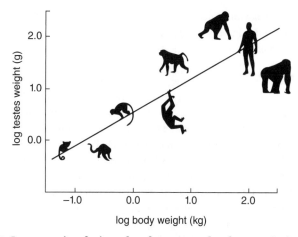

18. **Larger species of primate have larger testes, but those species in which females mate with multiple males have males with larger testes for their body size than do species in which females mate with only a single male.**

might expect, then, bonobos have the largest testes for their body size of all primates. Humans are socially monogamous and fall somewhere between gorillas and bonobos. While we may not be as free of the selection pressures of sperm competition as gorillas seem to be, sperm competition was probably not as important in our evolutionary history as it has been for bonobos.

Similar patterns of testes size variation have been reported from almost every animal group that has been examined. For example, in species of birds where females attend leks to choose among potential mates, they typically mate with just one or very few males (see Chapter 2 for more details). The males of lekking species have small testes relative to body size compared with the males of species such as jacanas where polyandrous females visit many different males to mate and leave their eggs for males to incubate. Species of insects where females mate with many males have relatively larger testes than monogamous species, and group-spawning fishes have relatively larger testes than solitary-spawning species. So ubiquitous is this finding that relative testes mass is now widely recognized as a measure of the strength of selection from sperm competition.

Variation in testes size among populations of the same species has also been linked to variation in selection from sperm competition. In the south-west of Western Australia during the rainy months of July to September, quacking frogs form choruses on granite outcrops in the forest where they defend shallow pools to which females come to spawn. Males have highly developed forearms and wrestle with each other for the possession of pools. Males with muscular arms are more likely to win these wrestling matches and enjoy a greater probability of mating than their weaker conspecifics, or at least they do in low-density populations. As the density of males in the chorus increases, males are unable to control access to females. In high-density choruses, shortly after a pool-holding male grasps a female for mating, other males will attempt to do the same, with up to nine males shedding sperm onto the female's clutch so that sperm must compete for fertilization (Figure 19). Because investment in forearm musculature trades off against investment in testes mass, males in a given population will emphasize one or the other, depending on how many competitor males are likely to occur.

Experimental evolution offers a powerful means for examining cause and effect. Given a suitable organism, one that has a relatively short generation time and can be reared in large numbers under controlled laboratory conditions, genetically isolated populations can be established and the strength of sexual selection manipulated experimentally over multiple generations before assaying the populations for traits such as testes size, sperm numbers, or competitive fertilization success. Such studies often use fruit flies of the genus *Drosophila*, but studies of other invertebrates and even house mice have been conducted. These studies have demonstrated that populations evolving under reduced post-mating sexual selection evolve decreased male

19. In Australian quacking frogs, several males can grasp a single female so that the ejaculates from multiple males must compete for fertilization of the female's eggs.

expenditure on sperm production while those evolving under increased levels of post-mating sexual selection evolve increased male expenditure on sperm production. These evolutionary changes in male expenditure can occur rapidly, within ten generations of experimental evolution, and provide direct evidence that post-mating sexual selection drives the evolution of male expenditure on sperm production, which then means they expend less on the weapons of pre-mating sexual selection.

Strategic ejaculation

Bateman argued that males should be limited in their reproductive success only by the number of females they can gain access to, assuming that sperm are in limitless supply (Chapter 4). However, we now know that this is not the case. Studies of insects, amphibians, reptiles, birds, and mammals including humans all show that males become depleted of sperm and seminal fluid following mating, and can take hours to days to recover their pre-mating supplies. We have already seen how male expenditure on the ejaculate may come at a cost to their expenditure on weapons and ornaments for mate acquisition. Dietary limitation can also affect the number of sperm produced and their quality, and sperm production can even reduce lifespan. In the Australian marsupials in the genus *Antechinus*, male expenditure on sperm production is particularly extreme. The mating season of this small mammal is very short and highly synchronized with the availability of their insect diet. Females can mate with many different males, and male expenditure on the ejaculate is characteristically high. Remarkably, males invest so much on sperm production that they suffer immune function collapse and death, leaving a population of single mothers. Given that males have a limited supply of sperm, sperm competition games predict that it would be advantageous for them to deliver their sperm prudently depending on the current risk of sperm competition. That is, when females are plentiful and rival males rare, males are expected to transfer just enough sperm to ensure fertilization and conserve their sperm supplies to increase the number of females they can inseminate. However, when females are rare and competition for access to them intense, males should deliver high numbers of sperm to compete for fertilizations.

Males of many species have been found to be acutely sensitive to the risk of sperm competition from rivals, and to increase the numbers of sperm they transfer to females when risks are high. This flexibility was first studied in insects by Matthew Gage and his colleagues, who counted the numbers of sperm transferred when a male Mediterranean fruit fly or mealworm beetle copulated in the presence of a rival male, finding that when a pair were accompanied by a rival the copulating male deposited twice as many sperm as when the pair were alone. This work has been extended to a host of other insect species as well as to birds, fishes, reptiles, and mammals.

Males can use a variety of cues in their environment to assess sperm competition risk, and do not rely simply on the presence of rivals. For example, male bank voles exposed to the odours of rivals and field crickets exposed to male songs both increase their expenditure on the ejaculate, and work with fruit flies suggests that although single cues in isolation are insufficient to affect adjustments in male ejaculate expenditure, any two cues from sound, smell, or touch do elicit male responses. Thus, males appear able to assess their social environment in such a way as to engage in sperm competition for fertilizations when necessary, while reserving their sperm for additional mating opportunities when rivals are scarce.

Avoiding sperm competition

Parker recognized that sperm competition would favour opposing adaptations in males that allowed them to gain precedence over any sperm stored by females from previous males, while at the same time preventing their own sperm from being supplanted by rivals. How does that happen?

Perhaps the best studied cases of such adaptation to sperm precedence come from the damselflies and dragonflies. Males typically search for females around the perimeters of ponds and streams where the females come to lay eggs. They engage females in tandem by grasping them at the back of the head with claspers located on the final segment of the abdomen. The males' primary genitalia are also positioned on the last abdominal segment, but secondary genitalia are located on the ventral region of the thorax. The male ejaculates into his secondary genital cavity before grasping the female, who will then reach under with her abdomen to engage the male's secondary genitalia in the so-called wheel position (Figure 20). The male then uses his secondary genitalia to physically remove sperm of any previous males. To facilitate this process, the 'penis' is endowed with backward-facing spines that

20. Damselflies are famed for their ability to remove sperm from the female's sperm storage organ before they deliver their own ejaculate. While in copula (a) the male uses his secondary genitalia (b) to entrap and scoop out sperm of rival males. In this way the mating male can monopolize paternity.

entrap sperm and drag them out of the female's reproductive tract. Only when cleared of rival sperm does the male deliver his own ejaculate. The result is almost complete paternity of any offspring subsequently produced. By removing sperm from rivals, males can effectively avoid sperm competition. The problem, of course, is that once mating is terminated the male risks his own sperm being displaced by any male that may encounter the female before she has produced her offspring. Males attempt to solve this problem by staying close to a female after mating, a behaviour called mate guarding.

Mate guarding is a conspicuous feature of animal mating behaviour. In damselflies, after the female disengages her genitalia, the male will retain his hold on her and fly in tandem with her to areas of aquatic vegetation where she will lay her eggs. The female is only released once she has deposited her full clutch of eggs. In many

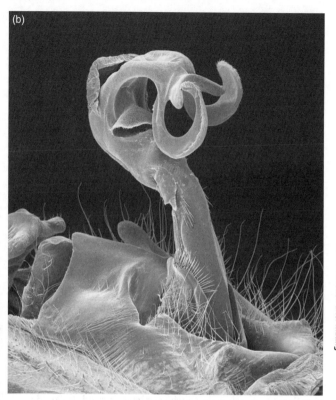

20. The secondary genitalia of a male damselfly.

species of insects, copulation is prolonged well after sperm have been transferred, to prevent rival males from copulating. Such copulatory mate guarding can last for days or even weeks, even though sperm are transferred in the first seconds or minutes of copulation. For birds, once an egg is ovulated it is available for fertilization for 4–5 days before it is bound in its shell and laid. There is a window of opportunity, starting some days before the first egg is laid and lasting up until the laying of the penultimate egg, in which an insemination has a chance of resulting in a fertilization, and it is during this time that the risk of lost paternity for males is at its greatest. During this period, male birds will typically maintain a larger territory about their nests, will increase the frequency of copulations with their partner, will have greater rates of singing activity to repel rivals, and will generally remain close to their partner until the final egg is laid.

Studies in human psychology suggest that mate guarding may even work in our own species to increase the likelihood of paternity. The use of 'menstrual huts' by the Dogon people of Mali appears to allow men to monitor their partners' fertility cycles and significantly reduces their risk of cuckoldry. More generally, men's use of behaviours such as controlling women's activities, behaving aggressively toward their partner's male friends, or other behaviours that are broadly characterized as sexual jealousy, have been argued by evolutionary psychologist David Buss to serve a mate guarding function. Certainly these male behaviours seem widespread, being reported in all but four of 849 human societies.

Despite its obvious benefits in avoiding sperm competition, mate guarding can also incur significant costs. While males guard their current mate, they are unable to search for additional females or defend their territories. In birds, males can experience significant weight loss during the egg laying period. It would pay males, then, to adjust their mate guarding activities depending on the immediate risk of sperm competition. Indeed, mate guarding by male birds is most intense in the days leading up to egg-laying. Among insects that practice copulatory mate guarding, the duration of extended copulation depends on the local sex ratio. Thus, when males are rare and females plentiful, males terminate mating associations soon after sperm transfer. However, when females are rare and rival males abundant, copulatory mate guarding is extended to hours or days after insemination. And in some species, males have evolved ways to prevent rival males from copulating without remaining with their mates. Thus in many insects, reptiles, and mammals, males deposit mating plugs in the female reproductive tract, sticky substances that function rather like the stopper in a drain. These can then block access to the female's reproductive tract by rivals. Male *Heliconius* butterflies deliver chemical compounds that render females unattractive to rivals, while in a variety of insects males deliver compounds in the seminal fluid that affect the female's physiology so as to make her unresponsive to the mating attempts of subsequent males. In many species such as fruit flies and katydids these receptivity inhibiting seminal fluid compounds can be short lived, while in other species they can have permanent effects on female behaviour, resulting in lifelong monogamy.

Cryptic female choice

Females that mate with more than one male are not simply providing an opportunity for sperm competition—they may also benefit themselves from the presence of sperm from multiple mates in their reproductive tract. We discussed in Chapter 3 the so-called 'sexy sons' and 'good genes' models of female preference evolution in the context of pre-mating sexual selection acting on male ornaments. These concepts have been extended to the post-mating arena to explain the evolution of polyandry. Under the so called 'sexy sperm' hypothesis, polyandry is beneficial for females because the traits in males that make them successful in sperm competition are heritable. Thus by mating with multiple males, females can produce sons that are superior sperm competitors simply because the sons inherit their fathers' abilities to win at fertilization. This idea is extended under the 'good sperm' hypothesis, whereby competitive fertilization success is proposed to be genetically linked to the general health and viability of an individual. Males that are successful in sperm competition produce not only sons who are successful in sperm competition, but also sons and daughters with greater general health and viability. For example, in the case of *Antechinus* we have seen how extreme multiple mating by females drives suicidal expenditure on sperm production by males. In the brown antechinus, females can gain important fitness benefits from promoting this intense sperm competition because males that win paternity sire offspring that are three times more likely to survive to weaning. Thus by promoting sperm competition females can obtain genetic benefits for their offspring in much the same way as pre-mating female choice based on male ornaments.

Females can play an even more active role in determining the outcome of sperm competition. Ultimately it is the female that has the final say over who will fertilize her ova, and biologist William Eberhard has highlighted over twenty different mechanisms by which they may do so. For example, females can choose to accept, store, nurture, and utilize sperm based on characteristics of prospective fathers. They can determine the number of eggs they release and fertilize, or the amount of resources they provide to those eggs or resultant offspring based on the characteristics of their fathers. Each of these mechanisms of so-called cryptic female choice can impose selection on male traits that females find attractive and/or offer the greatest reproductive returns for their investment. It's important to note that this 'choice' is not at all conscious—females of any species, insect or not, cannot exert conscious control over the operation of their reproductive organs. But if some sperm are favoured over others, the result is the same as if they could.

In some cases the traits subject to cryptic female choice can be the same traits that are subject to pre-mating female choice. We have seen in Chapter 3 how female guppies prefer males with the most orange coloration, which means that more orange males should tend to father more offspring. Using experimental manipulations of a female's perception of male colour, Andrea Pilastro and his colleagues have shown that female guppies are more likely to retain sperm when exposed to more colourful males, even though these were not the males actually mating. Moreover, even when interactions between males and females are circumvented by artificial insemination, the sperm from males with more orange coloration are more likely to gain paternity. Male guppies are orange, then, because of selection both before and after mating.

Male secondary sexual traits are typically highly divergent among animal species, a pattern ascribed to sexual selection and one that was discussed in detail in Chapter 3. Animal genitalia show this same pattern of divergent evolution, and William Eberhard has championed the idea that sexual selection via cryptic female choice is responsible for male genital evolution (Figure 21). In particular, it is not genitalia per se that exhibit patterns of divergent evolution, but rather any structure that makes intimate contact with females during copulation. For example, structures that are inserted into the female, such as the damselflies' secondary genitalia discussed previously, would qualify, as would their anal claspers that grasp the female in copula. It was Jonathan Waage who first discovered that the damselfly penis serves the dual role of sperm removal and delivery discussed above, ascribing its evolution to selection from sperm competition. But we now know that cryptic female choice is also involved because it is stimulation of the female reproductive tract by the penis that results in sperm being released from the female's storage organ so that it can then be removed. Thus Eberhard has argued that stimulation of the female during copulation, so-called copulatory courtship, serves as a selective mechanism of cryptic female choice driving the evolution of male genital morphology.

The idea that sexual selection could affect the evolution of animal genitalia seems reasonable, but where is the evidence? Support comes largely from studies of insects, perhaps because this group already shows an astonishing amount of variation in sex organ size and shape. That variation, it turns out, is not random; it can predict variation in insemination and fertilization success, so that insect species subject to greater sexual selection have been found to have more complex genital morphology.

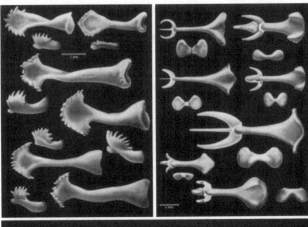

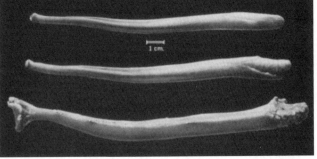

21. The mammalian penis bone or baculum is the most divergently shaped bone in the mammalian body. It is thought that sexual selection has played a significant role in this diversification. Top left, bacula of ground squirrels (*Spermophilus* sp.) have spoon-like distal ends with tooth-like projections that protrude from the glans penis; top right, bacula of rice rats (*Oryzomys* sp.) and voles (*Microtus* sp.); bottom, bacula of two species of bear (*Ursus*) above a sea lion (*Zalophus*).

Even more evidence that sexual selection can drive the evolution of male genitalia comes from studies using experimental evolution—placing laboratory animals under conditions that mimic natural or sexual selection. Populations of insects and house mice evolving under sexual selection, in which individuals competed for mates, or under enforced monogamy, have shown divergence in the shape of male genitalia. What about females? We should also expect to see similar patterns of divergence in female genitalia because female genitalia exert cryptic female choice on males. The few studies that have examined female genital structures have reported striking patterns of coevolutionary divergence between male and female genital traits, in animals ranging from beetles to fish and ducks, consistent with the notion that genital morphology is under cryptic female choice. Male genitalia seem to be evolving in a manner similar to the weapons and ornaments of pre-mating sexual selection, clouding the distinction between primary and secondary sexual traits.

We don't generally think of birds as having a penis, and indeed most birds do not. However, ducks are an exception. The Argentinian duck has a penis as long as its body, almost half a metre in length and shaped like a corkscrew. Duck penis length varies among species, and importantly covaries with female reproductive tract morphology as would be expected if cryptic female choice exerts selection for genital elaboration in males. However, female reproductive tracts are remarkably complex, having dead end sacs and clockwise coils. Interestingly, the female reproductive tract coils in the opposite direction to the penis, making forced copulation by males difficult. These findings suggest that the coevolution of male and female genital traits among duck species is driven by sexual conflict, a topic we will discuss in Chapter 6.

Chapter 6
Sexual conflict

Mallard ducks form monogamous pair bonds in spring, with females choosing among available males based on the yellowness of the bill and the brightness of feather ornamentation. These male secondary sexual traits appear to convey information to females on mate quality, including resistance to disease and fertility. Female choice in these ducks thereby affords females direct, and possibly indirect, benefits for the offspring they later produce (Chapter 3). Once a female has chosen a male she will resist the advances of additional males. However, males frequently attempt to force copulations on females (Figure 22). Forced copulations are costly for females, not only because they can result in offspring sired by non-preferred males of lower quality, circumventing female choice, but also because they can incur physical damage or the transmission of disease. Not surprisingly, females strongly resist forced copulations. But of course even a forced copulation can increase a male's fitness. Mallards offer a stark illustration of the conflict of interest between males and females over mating. While male reproductive interests are served by mating with as many females as they can, female interests are usually best served by mating with only the best available male.

Mallards, like other ducks, have an elongated penis that undergoes explosive erection. The Muscovy duck penis can be erected in 0.36 seconds, achieving an astonishing maximum

22. A group of male mallards forcing copulation on a single female who is being pushed below the surface of the water with considerable risk of drowning.

velocity of 1.6 ms^{-1}. This remarkable structure appears to be an adaptation in male ducks for forced copulation. In turn, the costs of forced copulation for female ducks appear to have favoured the evolution of genital tracts that resist such copulation. As discussed in Chapter 5, the structure of duck genitalia marks the battleground of evolutionary conflict.

People often joke about the battle of the sexes, or bemoan the lack of understanding between men and women. Does the idea of males and females being at odds have any basis in biology? Yes and no. The reproductive interests of males and females will almost always differ, for example over whether to mate and how often, when to produce offspring and how many, or how much to invest in each offspring. Whenever the reproductive interests of males and females differ, opposing selection on males and females to achieve their preferred outcome will generate sexually antagonistic selection. Such sexual conflict is reflected in differences in the appearance and behaviour of the sexes as each evolves to gain the advantage in a fitness 'arms race'. In this chapter we shall explore the evolutionary consequences of these arms races in the context of sexual selection as it occurs both before and after mating.

Chase-away sexual selection

Traditional sexual selection theory proposes that females prefer more ornamented males because the material or genetic benefits signalled by those ornaments mean that the choosy female's offspring are better off than the offspring of less discriminating females. This in turn causes selection for stronger preferences. But where do the male secondary sexual traits come from to begin with? One possible origin of such traits discussed in Chapter 3 revolves around the notion that female senses may be pre-adapted to favour novel male traits in a sexual context. For example, say that females are good at seeing the colour red because they often eat red foods and selection has favoured the ability to pick out the choicest berries. A male with a red patch on his chest will be more visible to females than one with a blue or green patch, which then gives him—and his ornament—a head start in selection for redness. Males get redder and redder over evolutionary time, and redness could even become an indication of a male's overall health and vigour (Chapter 3).

At the same time, however, such sensory exploitation by males may result in females mating at times when they should be foraging, and thus reduce their ability to gather fruit, or result in them mating more frequently than is optimal for their reproductive success. The costs associated with females responding to male redness in a sexual context might then favour a reduction in their sensory threshold to red. Females essentially evolve resistance to the male sexual trait, returning their mating

Reduced female
fitness

Trait increases male
fitness through mating
and/or fertilization
success
[Persistence]

Selection for decreased
female response to
male trait
[Resistance]

Selection for
increased trait
expression in male

23. **Chase-away sexual selection. Coevolutionary cycles of increasing resistance to traits that increase male fitness at a cost to female fitness. Such traits might be ornaments that stimulate females to mate more than is beneficial for them, or genital and/or ejaculate traits that affect fertilization success of males while reducing female lifespan.**

behaviour to its optimal level. In turn, males will be selected to increase the intensity of red coloration to meet the females' reduced threshold, generating cycles of opposing selection for female resistance and male persistence (Figure 23). A point might be reached when further reductions in female sensory threshold to red are countered by natural selection acting against females no longer able to find fruit. This process is called chase-away sexual selection. It differs from the other models for the evolution of mate choice we have discussed because in the chase-away scenario, female preference for a male trait actually decreases rather than increases, due to the costs of mating more often than is ideal for the females. Selection is still expected to act on males that have the traits females perceive the best, though, with the most obvious traits being those already subject to strong natural selection in females, such as those for predator avoidance or foraging.

For example, bats are a major predator of moths, locating their prey using ultrasonic echolocation. In turn, many moth species have evolved hearing mechanisms tuned to the ultrasonic frequencies of echolocating bats. When a moth hears the call of a bat it will become instantly immobile, falling from the sky to avoid being eaten. The males of several moth species, such as the Asian corn borer, produce courtship calls consisting of high frequency sounds that act in a manner similar to the ultrasonic calls of bats; on hearing a male's courtship call the female becomes instantly immobile, which allows the male to mate with her. Female Asian corn borers are unable to distinguish between the calls of bats and the courtship calls of males, responding in an identical manner to both, suggesting that male sexual calls have arisen to exploit the naturally selected auditory system of females. This type of sensory exploitation is not limited to moths; it has recently been reported in some types of crickets.

We saw in Chapter 3 how male water mites exploit the predatory response of females during courtship to facilitate mating. In one group of freshwater tropical fish, the characin or tetras, several species have independently evolved male secondary sexual traits that function as 'lures' to attract females. These 'lures' consist of extensions to the gill covers or modified scales on the flanks that are twitched and waggled in front of the female, who chases and nips at them as if they were a prey item. Once the male has a 'bite' he will rush at the female, depositing sperm in her genital opening. These examples offer compelling evidence for a sensory exploitation origin of male secondary sexual traits.

Male dance flies, small to medium-sized flies that prey on other arthropods, meet females in mating swarms. The males enter a swarm carrying a prey item, which is offered to a female to eat during copulation. In some species of dance fly it is the females that bear exaggerated secondary sexual traits in the form of extended leg scales and inflatable abdominal segments that exaggerate their visual appearance of size. Why the reversal of the usual pattern? One interpretation is that females exploit the visual system of mate-searching males to acquire the food gifts that males provide, perhaps forcing males to allocate their mating resources in a suboptimal manner. However, at least in one species of dance fly, females with larger ornaments lay more eggs, so that male preference for more exaggerated female ornaments also helps them in a more direct way.

It is extremely difficult to determine whether preferences evolved because of their benefit to females or because of sexual conflict. One way to distinguish between the alternatives, at least in theory, is to examine the economics of mating. Sexually antagonistic selection, and hence sexual conflict, arises because of the costs associated with mating with attractive males, while other sexual selection processes rely on benefits accruing to females from attractive males. In reality, trying to distinguish between coevolutionary mechanisms is probably a futile endeavour because one is unlikely to proceed in isolation of the other. There will always be a degree of sexual conflict between males and females, even in a system dominated by female choice for males offering material resources or good-genes to females, because counter to their fitness interests, some males will be rejected during the mating process. Nevertheless, in some cases sexual conflict has clearly been demonstrated to have contributed to the evolution of male and female reproductive traits.

Conflict over mating

Water striders are semi-aquatic insects that can be found skimming over the surfaces of ponds and streams where they forage for invertebrate prey. Water striders, unlike many animals, have no courtship rituals, and there are no secondary sexual traits associated with mating in these insects. Rather, males simply chase and seize females when they are encountered, and attempt to mate with them. Females will flee from males and if seized will resist copulation. Intense struggles ensue, during which the pair flips and summersaults as the female tries to dislodge her unwanted suitor.

Water striders are one group in which the economics of mating—in other words, the costs and benefits to each party—are well understood. During mating and subsequent mate guarding (Chapter 5) the female must carry the male on her back, which is a heavy burden to bear both because it literally increases the energy she expends moving on the water and because it decreases her foraging efficiency. Moreover, females run a higher risk of being spotted and eaten by a predator during mating and mate guarding. But resisting copulation is not without cost, either; mating struggles also increase energy consumption and exposure to predation. Female water striders tread a fine line to minimize these costs, and studies show that they are more willing to mate when the costs of resistance exceed the cost of mating.

Male water striders use their clasping genitalia when they attempt to get females to mate. However, females have spines on their abdomen that make clasping difficult (Figure 24). Göran Arnqvist and Locke Rowe have shown how these structures in the two sexes are evolving under conflicting selection pressures in this group of insects. Who wins the race, males or females? We can answer this question by looking at the evolutionary tree for the group, which shows us not only which species have a most recent common ancestor, but when a particular trait arose. Among water strider species there is no relationship between the degree of male and female armament and the outcome of pre-mating struggles. This is to be expected where male adaptation and female counter-adaptation keep females at their optimal mating frequency (Figure 24). However, differences in the relative armaments of males and females within species do predict the outcome of struggles. In species where male clasping genitalia are more exaggerated than female anti-clasping structures, mating struggles are more prolonged and males are more likely to succeed in mating. In contrast, in species where female anti-clasping structures are more exaggerated than male clasping structures, mating struggles are shorter and males less likely to achieve mating (Figure 24). This example shows us how

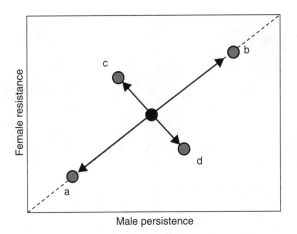

24. **Sexually antagonistic coevolution will not be readily apparent when observing the outcome of male–female interactions. As male persistence increases (b) or decreases (a) female resistance is expected to follow so that the outcome of interactions remains unchanged. However, sexual antagonism will be revealed when females (c) or males (d) are ahead in the arms race. Such imbalances can occur naturally as seen among species of water strider, or can be engineered in order to test sexual conflict theory.**

hard it can be to detect sexually antagonistic coevolution; if one sex evolves an adaptation, the other sex evolves a counter-adaptation to overcome it, and hence we only see the evidence of conflict when one sex is well ahead in the arms race.

We can also see evidence of sexual conflict if we examine when in evolutionary history the traits involved in mating struggles arose. In diving beetles, males possess suction cups on their front legs that allow them to attain a firm grip on the dorsal surface of the females' wing covers. A male's grip is so strong he can hold four times the weight of a female. Females will dive repeatedly when grasped, in order to dislodge males and prevent them from mating. Within many species of diving beetles, females from different populations show different wing forms that are better or

25. Suction cups on the protarsus (top left) and mesotarsus (bottom left) of the diving beetle *Graphoderus zonatus verricifer* are used to grasp females prior to mating. Some females possess roughened wing cases (middle) compared with the typical smooth wing cases (right) that prevent males from establishing a position.

worse at deterring male attachment: some might have smooth wing covers, while others have granulated, grooved, or roughened wing covers that prevent males from gaining suction (Figure 25). Among populations of one such species, the spangled diving beetle, the females with such attachment-resistant wings are more common when males in the same population have more and better suction cups. This association suggests that these populations are evolving under a sexual arms race. Perhaps most interesting, a single origin of male suction cups shows up on the evolutionary family tree of the group, and different kinds of resistant female wing covers have evolved independently five separate times, all of them after the male suction cups evolved. The sexually antagonistic arms race, like all arms races, is never-ending; remarkably, among some species of diving beetle, counter-adaptations in the structure of suction cups have arisen to deal with the particular type of roughening morphology in females.

Traumatic insemination is perhaps one of the most extreme examples of sexual conflict over mating. In bed bugs and plant

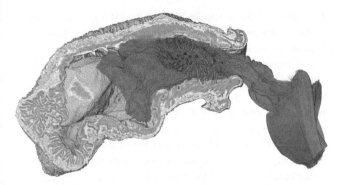

26. MicroCT scan of the aedeagus of *Callosobruchus maculatus* within the female reproductive tract. The spines in the mid-region of the aedeagus penetrate the walls of the female reproductive tract causing considerable damage.

bugs, rather than struggle with females over mating, males bypass the female reproductive tract and use their hypodermic-like genitalia to simply stab females in the abdomen and ejaculate into the body cavity. Traumatic insemination is of course very costly for females, increasing the potential for disease transmission and reducing their lifespan. As with the insects discussed in this chapter, in both of these groups of bugs we can see evolutionary evidence of counter-adaptations in females to resist the harm caused by traumatic insemination. The females of some species have structures at the site where males stab them that serve to receive the male's genitalia and minimize the degree of damage caused.

Less dramatic cases of traumatic insemination are also known. For example, in seed beetles of the genus *Callosobruchus*, the male's penis is endowed with spines that penetrate the internal reproductive tract of the female (Figure 26). Like bed bugs and plant bugs, the harm inflicted on females during mating in these beetles can reduce female lifespan, and females violently kick males to terminate copulation. We can see a signature of this conflict over mating by comparing male and female reproductive structures in different species of the beetles. It turns out that the length of male genital spines is positively associated with the thickness of the female reproductive tract walls. Why males should damage females in this way is not yet clear. It may be that damaging females has no direct fitness benefit for males, but arises simply as a by-product of sexual selection acting on males to increase their reproductive success. Recent work suggests that by piercing the reproductive tract walls seminal fluid proteins can be delivered to the female's bloodstream and in some way influence a male's competitive fertilization success. The harmful effect on female longevity might thus be collateral damage.

Sexual conflict after mating

When females mate with more than one male we expect selection to favour traits in males that promote their fertilization success over that of their rivals (Chapter 5). Sperm competition itself might generate sexually antagonistic coevolution, as hinted at in the beetles discussed here.

Sexual conflict after mating is perhaps best studied in the humble fruit fly, *Drosophila melanogaster*. In these insects, males include a cocktail of proteins along with their sperm, and those proteins affect their competitive fertilization success (Figure 27). Some of the proteins interfere with rival sperm in the female's sperm storage organs, facilitating displacement by sperm of the mating male. Other seminal fluid proteins, notably one called *sex peptide*, enter the female's bloodstream and from there travel to the ovaries to stimulate an increased investment in egg production and oviposition, and to the female's central nervous system, where they trigger neural and hormonal responses that make her unreceptive to the courtship attempts of rival males.

The benefits of these seminal fluid proteins for male reproduction are clear. The female will increase her production of offspring fathered by the copulating male while resisting copulations from

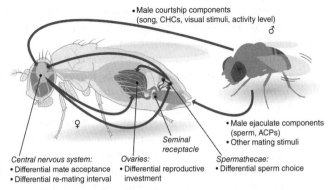

• Male courtship components
(song, CHCs, visual stimuli, activity level)

♂

♀

*Seminal
receptacle*

• Male ejaculate components
(sperm, ACPs)
• Other mating stimuli

Central nervous system:
• Differential mate acceptance
• Differential re-mating interval

Ovaries:
• Differential reproductive
investment

Spermathecae:
• Differential sperm choice

27. **Male fruit flies, *Drosophila melanogaster*, transfer a cocktail of accessory gland proteins (ACPs) in the seminal fluid that affect female reproductive physiology. One protein, sex peptide, affects pathways in the central nervous system that regulate female receptivity to rivals and has the side effect of reducing female lifespan.**

other males. But for females, these proteins can be harmful; *sex peptide*, for instance, reduces female lifespan. One way to see how sexual conflict has affected evolution is by examining the evolutionary tree and looking for variations in male and female structures associated with mating, as illustrated by some of the studies discussed in this chapter. Another way is to simply recreate the evolutionary processes that are suspected to lead to sexual conflict. William Rice did just that, by taking populations of flies and preventing the females from evolving in response to male harm. After many generations, males became even better at sperm competition, but females suffered higher mortality. Interestingly, female counter-adaptation to male harm in these flies uses proteins that directly attack the substances in the seminal fluid that harm the females. In these flies, then, males and females are engaged in chemical warfare over when and with whom to reproduce, and the signature of this coevolution can be seen at the molecular level, for example in the structure of reproductive proteins. Similar patterns of reproductive protein evolution are emerging in a variety of animals as diverse as gastropods and mammals.

One of the most universal adaptations to sperm competition among males is the production and transfer of large numbers of highly motile sperm that are capable of fertilizing an egg (Chapter 5). Such large numbers of competitive sperm, though, may not be in the best interests of females, who require only a single sperm to fertilize an egg. Indeed, polyspermy, the entry of more than one sperm into an egg, is fatal in a number of animal groups, leading to death of the newly fertilized embryo. The very fact that ova are generally resistant to fertilization and rapidly generate blocks to sperm penetration following fertilization has thus been interpreted as a result of sexually antagonistic coevolution between male and female gametes. Testing this hypothesis is extremely difficult, though not impossible.

We can also use experimental evolution to determine who 'wins' when either males or females are given the advantage in an arms race. Such an approach has been used to study sexually antagonistic coevolution between sperm and eggs in house mice. When populations of mice were allowed to reproduce under enforced monogamy or polygamy for twenty-four generations, they evolved differences in the number and motility of sperm. Males from monogamous populations evolved to have fewer, less motile sperm, while males from polygamous populations evolved to have greater numbers of faster-swimming sperm, as would be expected from the differences in selection from sperm competition among these populations.

Using in vitro fertilization, the same kind of reproductive technology used to help infertile human couples, population crosses in the mice have uncovered evidence of sexually antagonistic coevolution over the ability of sperm to fertilize eggs and the eggs to resist the advances, so to speak, of the sperm (Figure 28). Thus, the proportion of eggs fertilized using in vitro fertilization was greater when the eggs were mixed with sperm from polygamous population males than when the eggs were

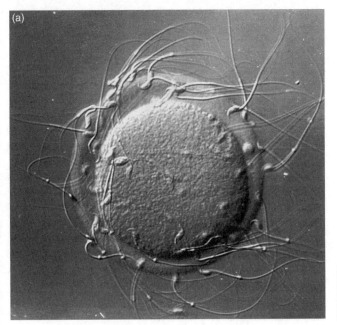

28. (a) Mouse sperm struggle to be the first to fertilize an ova.

mixed with sperm from monogamous population males. This
shows us that males in polygamous populations evolved higher
fertilization efficacy. On the other hand, eggs from polygamous
populations were less likely to be fertilized than those from
monogamous populations, suggesting that females from the
polygamous populations evolved resistance to fertilization. Finally,
the lowest proportion of eggs was fertilized when sperm from the
monogamous population was mixed with eggs from the
polygamous one, while the highest proportion fertilized was
between polygamous sperm and monogamous eggs. These
patterns support the idea that sexual conflict, fuelled by male
adaptation to sperm competition, favours the evolution of
defensive traits in females, or rather their ova.

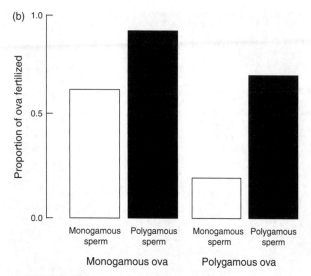

28. (b) By engineering an imbalance in the arms race between males and females with experimental evolution, sexually antagonistic coevolution between sperm fertilization competency and ova resistance to fertilization can be revealed.

How ubiquitous is sexually antagonistic coevolution?

As we have seen, sexually antagonistic coevolution can be a potent force in evolution, and can explain many behavioural, physiological, and morphological adaptations for reproduction in males and females. Sexual conflict was heralded as a 'new paradigm' in sexual selection, and existing knowledge was briskly reinterpreted in light of sexual conflict. It is even tempting to speculate that ideas about toxic sperm and other harm accruing to females as a consequence of mating fed into stereotypes about coy females that have lingered since Darwin. However, it would be wrong to conclude that reproduction is generally characterized by ongoing conflict. A few animal groups, like the *Drosophila* or seed beetles, certainly show evidence of such antagonistic coevolution, but evidence from other insects shows that seminal fluid proteins can also be beneficial to females, contributing to egg production and offspring quality, and even extending the lifespan of females in many species. Perhaps tales of embattled males and females were appealing to the popular imagination, while more pedestrian findings of sexual harmony drew less attention.

Parker's theoretical models of sexual conflict predicted evolutionarily stable resolutions in which either sex wins depending on the strength of selection on males and females and the starting conditions. This means that evolutionary arms races can often be avoided. For example, in Australian field crickets, seminal fluid proteins can reduce female lifespan. However, the benefits they bestow in the form of increased embryonic survival of offspring balance any costs of reduced lifespan, so that female lifetime reproductive success is unaffected. In this case the seminal fluid proteins of males do not conflict with female reproductive interests in any evolutionary way. The study of sexual conflict is still in its infancy. The literature is dominated by work on just three model systems, *Drosophila*, seed beetles, and water striders. We need much more research on a greater diversity of animal groups before we can comment on the broad evolutionary consequences of sexual conflict.

Chapter 7
How sex makes species survive

How new species come about is one of the most fundamental
questions in biology. And because that process involves reproduction,
and because reproduction often (though not always) means sexual
selection, it should come as no surprise that sexual selection has
often been thought to play a role in speciation, the way that new
species arise and are maintained. Indeed, Darwin noted that many
of the groups with the highest number of different species, such
as birds of paradise, were also those that had the most elaborately
ornamented males. As we pointed out in Chapter 1, sexual
selection, and the dizzying array of ornaments and weapons it
produces, is now recognized to be far more than simply a means
for females to recognize a male of the appropriate species. It can
also be the engine that drives new species to come about.

How do species form?

New species—which we shall define as groups of interbreeding
individuals that are reproductively isolated from other such
groups—arise when barriers to free interbreeding exist. Some of
those barriers are obvious: a mountain range divides populations
that used to be in contact, or shifting continents separate groups.
Some are less so, via changes in the timing when different
populations of insects visit plants for feeding and reproduction,

for example. Eventually, the populations become isolated and diverge into two separate species, with different genetic makeup.

If the populations encounter each other again, they may be insufficiently distinct and begin to interbreed, which means the speciation process was incomplete. If they cannot interbreed, barriers could exist at one of two levels. First, mating might not happen in the first place, perhaps because females do not recognize males as being appropriate mates, or because their genitalia don't quite fit together anymore. This is called pre-zygotic reproductive isolation, because the populations are separated before a zygote, or fertilized egg, has a chance to form. Second, even if a male and female from the two populations mate, a viable offspring may not be produced, because the genetic incompatibility between the two is too great. As we noted in Chapter 1, for example, mules are the offspring of a male donkey and a female horse, and while the mules themselves can survive, they cannot reproduce, and hence could not persist in the wild. Such separation is called post-zygotic reproductive isolation. Many studies claim that pre-zygotic isolation is more important to the speciation process than post-zygotic isolation, which means that sexual selection, the process crucial to mates getting together in the first place, is likely to play a role.

In either case, the two populations continue to evolve after they come into contact once again. One of the hallmarks of such contact is called reproductive character displacement, whereby certain characteristics used in mate choice evolve to become more different in zones of contact than they are in places where the two populations are unlikely to encounter each other. For example, different species of threespine stickleback, small fish that live in many parts of the world, sometimes inhabit the same lakes in Canada and sometimes occur by themselves. When two stickleback species occupy a single lake, they evolve two distinct forms. One feeds in open water and another feeds from the bottom of the lake.

The two types are different sizes and have different jaw and body shapes, all of which helps the two types eat different food. Hybrids between the two forms do not survive or reproduce as well as either parental type. If, however, only one species occurs in a lake, it has characteristics in between those found when multiple species co-occur. Character displacement has been demonstrated in a number of other animals, including frogs, which evolve differences in the calls used to attract mates when similar species overlap, and birds, which may have more pronounced bill differences and more distinct diets if they are in the same area as other closely related species.

Which is more important, food or sex?

The relative importance of natural and sexual selection in species formation has been hotly debated. Animals that come to live in different habitats undergo different kinds of natural selection, so that ecological factors clearly drive at least some population divergence. But as long as sexual selection makes some individuals more attractive as mates than others, the potential exists for it to also shape the way that populations interbreed and hence the likelihood of separation into different species. Ecological differences usually have to do with access to food sources, so that natural selection is synonymous with selection on foraging ability. But searching for mates is quite similar to searching for food, which makes the two processes similar. In a number of ways, then, sex is a lot like food, and both have the potential to take a single ancestral species and carve it into many distinct ones.

Imagine that a population of animals eats many different things, but every individual specializes on a particular food. How the individuals with those food specializations cluster as part of the larger population will determine how likely the population is to split up into different species. A similar pattern might occur with respect to mate choice, so that a population could include males

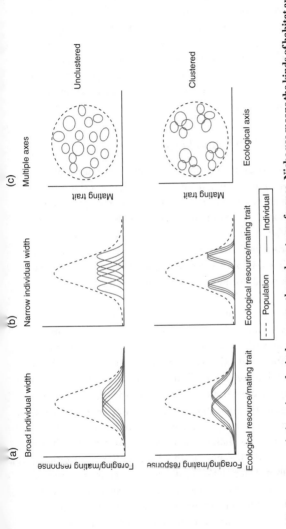

29. Pathways to speciation, via ecological means or through mate preference. Niche use means the kinds of habitat and food preferences an animal has—a fish that breeds in any body of fresh water would have a larger niche than one that only uses small rapidly running streams. More variation among individuals can occur either when an entire population shifts its preferences for mates or its niche use or when the individuals specialize within a population. Speciation is favoured so long as those differentiated individuals are more likely to mate with each other.

with a range of ornaments, like tails of different sizes. Different females, however, might prefer males with different tail lengths. In either case, diversification is facilitated when these individual preferences are narrow and clustered within a broader population preference (Figure 29). We still do not understand when, or why, such individual preferences for either food or mates cluster the way they do; such an understanding would go a long way towards helping us see how natural selection and sexual selection both act to make new species.

Toads, fish, locks, and keys

Some groups of animals provide particularly good illustrations of how populations separate into species because a single type of sexual signal is used to distinguish the appropriate partner for reproduction. Frogs, toads, and other animals that have mating calls are often used to study the effect of sexual selection on species formation. Their calls are easy for scientists to measure, record, and use in playback experiments, where they can see how females respond to different kinds of male signals.

Sexual selection was implicated in speciation in an Amazonian frog, *Physalaemus petersi*. Some populations of the frog produce a simple call with a component called a 'whine', but in other locations, males sometimes add another component, the 'chuck'. Females tend to prefer the calls from their home population, and even where the types overlap, there is little exchange of genes, suggesting that the populations are on their way towards separation. Even more telling, populations with different calls are more distinct genetically than populations that are separated by the same geographic distance but that have similar vocalizations. Presumably, the populations will become even more genetically distinct as time proceeds. The fact that a sexually selected trait diverged while the remaining characteristics stayed relatively similar among populations is further evidence that sexual selection itself drove the process.

Although conspicuous sexual signals like frog calls are important in distinguishing species from one another, recently scientists have realized that a less obvious characteristic can also be crucial in creating and maintaining the diversity of creatures on earth. In order for sperm and egg to meet, male and female genitalia have to be compatible. This would seem to be a relatively simple requirement, yet the genitalia of many animals, most notably insects, is astonishingly complex, with elaborate extensions and an assortment of spines and spikes in males and circuitous tunnels in females. These structures were originally thought to act as barriers to hybrids between different species in what is termed a 'lock and key' mechanism, in which only the appropriate 'key' from a male can be used to fertilize a female with the corresponding 'lock'.

In essence, the lock and key hypothesis is like the idea that sexual selection is all about species recognition. This idea, while intuitively appealing, does not explain the extraordinary diversity of genitalia among species, and it also does not explain why the complexity of genitalia varies with the mating system. What if, instead of merely functioning as puzzle parts that must fit together, genitals are subject to sexual selection via mate competition and mate choice, as other traits are? If sexual selection is responsible for the evolution of genital shape and size, we should expect to see patterns of divergence in genitals that correspond to the way sexual selection acts. Thus, in monogamous species, where females mate with only a single male, the opportunity for sexual selection to act on genital morphology during and after copulation is limited, so we should expect such species to exhibit far less evolutionary divergence in genital morphology than polyandrous groups. This expectation seems to be upheld in several different types of animals (Chapter 5).

As we have seen, natural selection can provide a check to the exaggeration of familiar secondary sexual characters such as songs or plumage if the ornaments make their bearers too conspicuous to predators. It turns out that similar trade-offs between sexual and natural selection occur with regard to genitalia, and may then generate rapid genital divergence and subsequent speciation. For example, male poeciliid fishes (the group that contains guppies and other aquarium pet species) possess a large, modified anal fin that serves as a 'penis' and can make it difficult for the fish to swim and avoid predators. This means that sexual and natural selection both act to change the appearance of the genitalia in such species. Indeed, males from populations of mosquitofish in places without predators have markedly different genital structures from those in places where predators are not present.

Trade-offs associated with sexual selection also play a role in the maintenance of separation of two species of spadefoot toad, desert-dwelling species whose tadpoles develop in temporary pools in the south-western United States. The two species will hybridize, even though the results of such cross-species matings are at a disadvantage: the hybrid males can be sterile, and the female hybrids produce fewer eggs than pure-species females. Why, then, do hybrids still persist? Wouldn't selection have removed them from the population?

The answer lies in the interaction between a toad's environment during its development from a tadpole into an adult and its species. If you are a tadpole in a rapidly drying pond, the faster you can develop into an air-breathing creature, the better off you will be. It turns out that the hybrid tadpoles develop faster than tadpoles from just one of the parental species. Depending on the conditions, then, a female from one species will actually have more surviving offspring if she prefers to mate with a male from a different species. And indeed, experiments have shown that female spadefoot toads prefer their own species' calls when the pond is deep and long-lasting, but a different species' calls when their pond is shallow and likely to dry up soon. In addition, female toads that are in good condition have offspring that develop faster. The preference for males of the 'wrong' species becomes more pronounced for females that are in poor condition, presumably

because they benefit by gaining any edge for their offspring
that they can, even if it means those offspring aren't as robust as
they could be.

Opposites rarely attract

Although the spadefoot toads illustrate a way that mating with
a different species can be advantageous, in most cases hybrids are
selected against because they are either less fertile or less able to
survive than either of their parental species. From an individual
animal's point of view, avoiding producing such hybrid offspring
in the first place would be ideal. One way to do that is to mate
with other individuals that are like oneself, so that, say, larger
males pair with larger females and smaller males with smaller
females. We call this process of mating based on an individual's
own characteristics assortative mating. Positive assortative mating
means that the members of a pair choose the same characteristics
as their own, while under negative assortative mating, they choose
opposite ones.

Although mate preferences are involved in assortative mating, it
is important to realize that it is distinct from sexual selection
and the two processes can occur separately or together. Sexual
selection occurs when individuals have differential reproduction
due to their ability to compete for mates, but assortative mating
may occur even if individuals all have equivalent reproductive
success; if all green males mate with green females and blue males
with blue females, but neither green nor blue individuals have
higher fitness, sexual selection is not acting. Conversely, if both
green and blue females prefer green males, assortative mating is
not apparent but sexual selection is operating because the green
males have higher reproductive success than the blue ones. It is
also worth noting that individuals of similar appearance can
end up being paired even though they are not using their own
characteristics as criteria. If high-quality individuals are able to
attract other high-quality individuals as mates, leaving the

low-quality individuals to choose among themselves, it may appear as though an individual chooses a partner on the basis of its own traits, when in fact, given the opportunity, everyone would prefer to have a high-quality mate. Similarly, if sizes of males and females covary over time and pairs are sampled over a long period, it may appear as if large and small individuals are selecting each other, but the available pool of mates is merely changing over time. Finally, assortative mating may occur if the act of courtship or copulation is difficult when partners differ substantially in size; constraints of this type have been suggested for frogs and certain kinds of arthropods, particularly aquatic species such as the water striders discussed in Chapter 6, in which one sex carries the other.

Assortative mating is seen in both invertebrates and vertebrates, including humans. In many cases, positive size-assortative mating occurs; for example, in many kinds of crustaceans and insects, large males are generally found coupled with large females. This pattern may occur because larger males are both more successful in mate competition and able to choose larger females that in turn lay more eggs.

Animals may also mate assortatively using behaviour rather than appearance. In many species of songbirds, females prefer to mate with males singing the same dialect of the song that their fathers sang. This type of pairing may occur through a process called sexual imprinting, in which young animals learn a set of characters during a critical period and then use those characters as criteria when finding a mate after they mature. Although sexual imprinting involves learning, it can have important genetic repercussions, because hybridization is discouraged and separation of populations may occur.

Among humans, positive assortative mating has been found for a wide variety of traits, including physical attributes such as height and relative weight as well as psychological ones such as a tendency toward antisocial behaviour. Environmental factors such as education level and socioeconomic status also seem to covary in members of couples, although in many societies a tendency for women to 'marry up' is seen, so that women are more likely than men to have partners who are better educated and from a higher social class than they are themselves. Interestingly, couples who are living together are at least sometimes found to be more different than married couples. Sociologists and psychologists point out that such covariation of non-genetic characteristics may obscure attempts to study heritability or the relative contributions of genes and the environment to the development of personality traits or such conditions as obesity.

Although positive assortative mating is more common, some fascinating examples of negative assortative mating occur. In the North American white-throated sparrow, two different morphs occur: a white-striped one, with brighter crown stripes, brighter yellow eye stripes, and less striping in the throat patch, and a tan-striped morph, which is also smaller and less aggressive (Figure 30). Both sexes exhibit the dimorphism, which is caused by variation in one of the chromosomes. Up to 98 per cent of the

30. In white-throated sparrows, both sexes appear in two morphs, with females preferring to mate with a male of the opposite morph from their own.

birds in a population mate with an individual of the opposite morph. When birds of both sexes and both morphs were presented with a choice between partners of each morph, both tan-striped and white-striped females preferred tan-striped males, at least so long as the birds were all in the same cage together and the males were able to interact with the females. If the females only observed their potential mates through a window, they exhibited no preference. Both types of males, on the other hand, preferred white-striped females, but only when they did not interact with the females. The negative assortative mating is maintained via the higher competitive ability of white-striped females, which are able to pair with the preferred tan-striped males.

Assortative mating might contribute to speciation in two contexts. It can cause a single population to split into two, resulting in speciation with the two groups still in the same place. Second, it could keep two populations that come into contact after separation genetically distinct, and thus prevent them from merging back into a single population. Because it is possible to have assortative mating without sexual selection, assortative mating does not always lead to reproductive isolation. Equally important is the degree to which females differ or vary in their mate preferences.

Sexual selection and the rate of evolutionary diversification

We know evolution happens to all living things, but why do some groups seem to have changed relatively little over long periods of time, while others split into many different species over tens or hundreds of thousands of years, a relative blink of an eye, evolutionarily speaking? Several groups of animals seem to have gone from a single common ancestor to a richly varied group of species particularly rapidly, and sexual selection is often implicated in the process.

The idea is simple: if different females have different preferences for male sexual signals such as complex songs or colourful plumage, and their offspring inherit those preferences, what was once a single interbreeding population could separate into multiple sub-populations, each with its own set of signals and preferences for them. If offspring tend to prefer mates that resemble their parents, this reinforces the separation. Over time, selection continues within each of the smaller groups, and they diverge even more, eventually producing a range of species.

Although this notion originated with Darwin, it is difficult to test directly. It is possible, though, to compare how many species are in a given group of animals and see if the tendency to diverge is correlated with the tendency to have extravagant sexual ornaments. One such study did just that, using songbirds from around the world. The researchers examined pairs of groups of birds, with the species within each group having arisen from a single common ancestor. Then they measured the degree to which males and females in the various species had different plumage. In some species, such as ovenbirds, the sexes appear almost identical, while in others, such as most tanagers, the males have brilliant colours and the females are relatively drab and inconspicuous. Presumably, sexual selection has been more intense in the species with larger differences between the sexes. The scientists then counted the number of species in each pair. Overall, the more different males and females within a group were, the more species that group had, which suggests that sexual selection has driven diversification in songbirds.

Similar processes have been suggested for a number of so-called *adaptive radiations*, the result of a rapid diversification into many species from a single ancestor when a change in the environment makes new resources available, or new opportunities for selection appear. Famous examples of adaptive radiations include the birds of paradise in New Guinea, the Galapagos finches studied by Darwin himself, the cichlid fish in Lake Malawi in Africa, and the lizards living on islands in the Caribbean. Island populations seem particularly prone to adaptive radiation, perhaps because once a colonizing individual arrives, many new opportunities and environments present themselves and its descendants can occupy a variety of habitats and take on a variety of different forms.

Some of the most dramatic examples of adaptive radiations involve animals much less conspicuous than birds of paradise or even lizards. The Hawaiian Islands have long been noted for their extraordinary diversity of animals and plants specialized to live only in a tiny area. In many cases, many different forms arose from a single ancestor that managed to make it to these isolated islands in the Pacific, and this is beautifully illustrated by the humble fruit fly. Most people think of fruit flies, in the family Drosophilidae, as rather dull tiny flies that buzz around rotting fruit or even rubbish heaps, but the Hawaiian Islands have over 800 different kinds of drosophilid flies, more than anywhere else in the world. All of the species have a single common ancestor, perhaps even a single female that floated or was carried on a bird's foot to Hawaii and then laid eggs in the new land. The flies we find in Hawaii today are strikingly different in their size and appearance, with body lengths of less than 1.5 millimetres (a sixteenth of an inch) to more than 20 millimetres (three-quarters of an inch). They occupy many different habitats, from lush rainforests in valleys separated by steep mountains to grassy fields at high altitudes (Figure 31).

The Hawaiian drosophilids also vary in their reproductive behaviour, with some laying just a single egg and others producing hundreds at a time. Sexual selection among the flies is equally variable, and differences in courtship and mating behaviour may be contributing to the continued speciation of this group. One species, called *Drosophila silvestris*, lives in cool, wet forests above 750 metres (2,500 feet) in elevation, laying its eggs in the decaying bark of trees. Many kinds of drosophilid flies have elaborate mating rituals, with males in some species producing songs of too

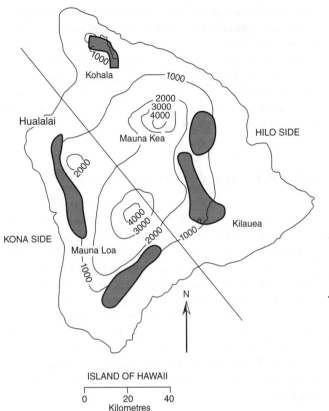

31. A map of the Big Island of Hawaii, with the five regions inhabited by different populations of the fruit fly *Drosophila silvestris* (shaded in grey). The populations on the Hilo side of the island, which are younger than the Kona side populations, are developing differences that over time could lead to the origin of a new species. The contours are in metres.

high a frequency for humans to hear, by buzzing their wings in complex patterns, and others perform dances that attract receptive females in the vicinity. The males of *D. silvestris* have a series of hairs on their forelegs that they brush against females during courtship, but they do not always use them the same way. On the north-eastern half of the island of Hawaii, called the Big Island, the males have many more of these hairs than do the males on the south-western side. Presumably, the difference evolved via sexual selection, and if females from the north-east continue to favour the kind of courtship that only a hairy-legged male can deliver, the populations will continue to differentiate in other ways that could eventually split them into two species.

Behaviour shaping species formation

The different species of drosophilids differ in both their appearance and behaviour, and while it has long been recognized that differences in the way animals look can lead to speciation by affecting how mates recognize each other, the role of behaviour in species formation has been more controversial. Although we know that behaviour in the form of courtship often mediates mate choice and hence the exchange of genes among different populations, it is not clear how the evolution of behaviour contributes to diversification. This is partly because behaviour is so flexible, changing during an individual's development or when animals are in different circumstances, and it is hard to understand the way that it can permanently affect the evolution of more tangible traits.

One study, though, did find that behaviour, at least in the form of bird song, was instrumental in species formation. While males of many birds sing, one major group, including the flycatchers and ovenbirds, only produces relatively simple songs with a few notes, while another, including thrushes and wrens, sings the melodious songs that inspire poetry and rely on a unique and highly specialized vocal apparatus to produce. The first group's songs are inherited, while the second group needs to learn its songs, usually from the father. Scientists examined the rates of speciation, determined using the genetic relationships among the members of a group, in a subset of both types of singers. They also estimated the evolution of songs among the various species, by comparing the components within each song and determining how much the songs changed over evolutionary time. In both groups, the higher the rate of speciation, the greater the rate of song evolution. Even more interesting, song evolution was more rapid in the group of birds that had learned rather than innate songs. This suggests that learning—at least of songs, and perhaps of other things as well—could make a difference in how fast a trait is acquired, and hence in the speed of evolution and speciation.

Another connection between behaviour, sexual selection, and the potential for changes in mate recognition comes from an examination of cognitive ability in guppies, not animals generally known to be intellectual powerhouses. Yet choosing a mate, regardless of the species, requires evaluation and complex decision-making, abilities that could drive diversification. Cognitive ability is linked to brain size, and scientists artificially selected for large-brained and small-brained guppies which duly demonstrated differences in problem-solving ability.

The researchers then predicted that the larger-brained female guppies should be better able to discern a high-quality, attractive male than their smaller-brained counterparts, and hence should more often choose an attractive male (as we saw in Chapter 3, a male with brighter orange spots). To test this prediction, they measured the preferences of females from each of the selected lines as well as those of females from wild-type populations that had not experienced any selection on brain size. Indeed, both the large-brained and wild-type females preferred the attractive males, but the small-brained females showed no preference for a particular male type. The difference in preference was not due to a difference in visual ability, which means that preferences for particular kinds of males could arise in various ways during brain evolution, and thus direct the way that populations become reproductively isolated.

Could sex prevent extinction?

Extinction could be viewed as the reverse of speciation, when instead of diversifying, populations disappear. Here too, sexual selection might play a role. Because it results in extreme traits like peacock trains, and because those traits are often detrimental or very costly to produce, in a few cases it has been suggested that sexual selection sometimes goes too far, so to speak, and can doom a species to extinction. The famous Irish elk is the best example of such apparent evolutionary over-indulgence (Figure 32). Also known as giant deer, this species was alive during the Pleistocene. Males were over 2 metres tall at the shoulders, and their enormous antlers were the largest of any member of the deer family, measuring up to 3.65 metres from one end to the other and weighing up to 40 kilograms. Fossil evidence suggests that the elk had a rutting season, much like modern deer, which means that

32. A male of the giant extinct Irish elk, shown next to a human skeleton for scale.

males probably competed for access to females. Antlers are quite costly to produce, and the severe competition for mates among males would probably have resulted in injury and sometimes death to the competitors. While we will never know for certain, many scientists have speculated that sexual selection for huge antlers eventually led to the Irish elk population having too few males to sustain itself, leading to eventual extinction.

On the other hand, sexual selection may be able to prevent extinction. If females prefer only certain types of males, and less-preferred males are less likely to mate, sexual selection provides a kind of filter that preserves only particular types of individuals. For instance, as discussed in Chapter 3, females could prefer males with fewer damaging mutations, so that mate choice 'clears the slate' of such less-favourable genetic variants every generation. Assuming that preferred males are of higher quality than less-preferred ones, this means that the overall fitness of populations subject to intense sexual selection would be higher than that of populations not experiencing such selection. The end result is a population less likely to go extinct.

To see if this idea is feasible, scientists performed a set of experiments using flour beetles, the small insects that, as their name suggests, infest flour and other stored grain products. The beetles were subject to either strong or weak sexual selection by varying the sex ratio, so that some populations had many more males than females (inducing intense mate competition), some had the reverse, and yet others experienced enforced monogamy, so that no mate choice (or sexual selection) was possible. The offspring from each population were then inbred (brothers were mated to their sisters) to expose any mutations that might have arisen over preceding generations. The populations were allowed to grow or shrink without any interference from the investigators. Indeed, the populations that had experienced strong sexual selection were more likely to survive inbreeding, and the scientists suggested that a history of mating competition could purge mutations from a population. Whether this means that sexual selection could help endangered species survive in the real world is still unknown, but the idea underscores how an understanding of sexual selection can have practical implications for conservation.

Chapter 8
Conclusions, and where to from here

We hold our ideas about sex and gender very close to our hearts, and almost nothing will guarantee an argument as much as a blanket generalization about male or female 'nature', at least as far as humans are concerned. Are we naturally monogamous? Are women more verbal, more emotional, or less inclined to be interested in mechanical tasks? Are men power-grabbing aggressors? And, key to our efforts in this book, what do our observations about animals tell us about the evolution of the sexes? Because sexual reproduction is central to the lives of most animals, writing about sexual selection therefore means writing about nearly everything in biology—about what animals look like, where they live, how they behave towards each other, and of course how the sexes are influenced by evolution.

We hope that the studies discussed in this book have underscored something that virtually all researchers in animal behaviour have come to realize: animals are far more variable in their sexual behaviour than is commonly believed, which means that they do not set a model for natural human behaviour, regardless of how we define it. Females are not always the more caring sex, males are not always more competitive than females, and mate choice and competition take many forms. Under the same selective pressures—to survive long enough to reproduce and to pass on genes to subsequent generations—animals have evolved many different solutions. Anglerfish males are passive sacks of sperm attached to a female, and female bush crickets prefer to mate with a male that presents them with a nutritious gift that weighs a third of his body mass. Male bedbugs bypass the female's genital tract, and instead stab their mate with a hypodermic penis, releasing sperm into her body cavity. Parasites may have played a role in the evolution of flashy male secondary sexual characters, so that female animals are acting almost like physicians performing a medical exam, scrutinizing their potential mates for signs of disease. Every day, it seems, researchers uncover a new and bizarre example of a trait used in sex. This dazzling variety also cautions us against relying too much on just one or a few model species to provide us with answers to how sexual selection works—it may be possible to derive our understanding of genetics using just fruit flies and bacteria, but such a limited set of examples will never work for sexual selection.

At the same time, the principles of sexual selection give us a framework in which to interpret all this variation. It would be wrong to throw up our hands and say that in the animal kingdom, anything goes. In *On the Origin of Species*, Charles Darwin famously said that 'If it could be proved that any part of the structure of any one species had been formed for the exclusive good of another species, it would annihilate my theory, for such could not have been produced through natural selection.' Similarly, we would argue that even the most outlandish of traits resulting from sexual selection can only evolve because they render their bearers more attractive or able to defeat their rivals, and hence increase their bearers' fitness in the context of sexual competition. Sometimes that fitness occurs at a cost from natural selection—colourful birds that sing loud songs may make themselves more conspicuous to predators, for example. But unless the benefits in terms of mate attraction or competitive ability outweigh the costs, the trait will not evolve.

Sexual selection theory also helps us interpret the broad patterns of sex differences we see across the different animal groups.

Why is it that in general, males rather than females tend to possess both the weapons used in fighting with other males of their species and the ornamental feathers, fur, and other traits that make so many species distinctive? The answer lies in how sexual selection acts on the two sexes. As we have seen, the ability of males to leave genes in the next generation is limited by the number of females they can inseminate. More mates—and the ability to compete for them—is the key ingredient for success. Females, on the other hand, have a limited supply of eggs, and those eggs can almost certainly become fertilized. While females too can benefit from multiple mating partners, their success depends largely on the quality of the few males they seek out as partners. The exceptions, like the competitive female and choosy male bush crickets, occur because in certain circumstances, males are the limiting commodity since they supply the costly nuptial gifts.

What are some of the newest findings in sexual selection research, and where is the field going? Sexual conflict has received a great deal of attention over the last decade or so, as Chapter 6 demonstrates. In the early days of animal behaviour research, it was thought that nothing could be more compatible than the interests of males and females in mating, since of course the ultimate function of producing offspring is the same for both sexes. Harmony was thus thought to be the rule, with reluctant females simply needing to be persuaded that the time was right by courtship and ornamentation. But theory about the often-opposing benefits of mating to the sexes has now shed light on why it often seems as if such harmony is not always realized. The idea that even the structure of the genitalia of males and females reflects opposing selection pressures on the two sexes reveals new ways to understand how and why males and females differ. But again, as with the influence of sexual selection more broadly, it can be tempting to generalize and over-emphasize such conflict, even as we see its effects on many behaviours and other traits. From an initial view that the sexes were always in synch to the more recent idea of mating as a battleground, we now hope the pendulum is equalizing. Mating, even multiple mating, does not necessarily harm females and benefit males, and conflict can appear with degrees of severity, ranging from nearly negligible to extreme.

Another relatively recent insight that continues to provide a rich new avenue for research is the recognition that sexual selection doesn't stop after mating. Both sperm competition and cryptic female choice have been demonstrated in a wide variety of species, though the former appears to be far more ubiquitous than the latter. Males invest not only in obvious weapons such as horns or large bodies, but in the size and structure of their sperm. And females can mate with several males but bias paternity towards only a subset of those males. These post-mating phenomena are often invisible to the observer, but their effects have important implications for evolution. One of the most powerful new tools to detect and evaluate sperm competition and cryptic female choice is experimental evolution, whereby researchers re-enact the processes of selection under carefully controlled conditions and witness the results in real time.

Genomics, the study of the entire set of genes in an organism, is offering us a way to understand all kinds of behaviour, including sexual behaviour, as never before. Using newly developed technology that can determine gene sequences for large stretches of DNA, scientists are able to ask questions about how the genes of an organism—its genotype—are reflected in how that organism looks and acts—its phenotype, a linkage that is particularly relevant to sexual selection. Because behaviour is so flexible, determining its genetic underpinnings helps us see how much selection is likely to be able to alter a given trait. Genomics also allows us to study not just the DNA, the so-called 'programme' for the genes, but the transcriptome, the RNA that tells us how those genes are being expressed. Having a gene is one thing; the real question is whether that gene actually shows effects in the organism that harbours it.

Nowhere has the application of genomics to animal behaviour in general been so successful as in its use with the social insects, the bees, wasps, and ants that exhibit extraordinary behaviours such as sterile workers that devote themselves to the welfare of their colony. How could such an unusual form develop? We have seen for hundreds of years how some honeybee larvae become workers while others become queens, but the genetic underpinnings were not clear until now. A variety of genomic tools have shown that many of the steps between egg and queen or worker are reversible, that genes are not a simple on/off switch for behaviour, and that similar pathways can be used to produce different outcomes.

More relevant to sexual selection is the use of genomics to understand the evolution of alternative reproductive behaviour, which we discussed in Chapter 4. For example, bulb mites are tiny arthropods related to spiders that inhabit, as their name suggests, a variety of bulbs, including flowers such as hyacinth or crocus, and even onions. One species of bulb mite has two heritable male reproductive types, fighters and scramblers, rather like the isopods mentioned in Chapter 4. Fighter males have a larger third pair of legs for fighting, while scramblers have legs similar to females and do not engage in fighting behaviours. In a study of the transcriptomes—the gene expression profiles—of both male reproductive types as well as of the females, genes more likely to be expressed in males were found to evolve faster than those genes more likely to be expressed in females. Furthermore, four times as many genes were biased towards expression in the fighter type of male as in the scrambler type. The scientists who studied the mites concluded that the difference between the male types was related not to whether or not a given gene was present, but to the way in which the same gene is expressed in different individuals.

Genomics has also been suggested as a useful tool for better understanding sexual conflict. Such conflict may even occur at a single location on a chromosome. In this case, a gene that when expressed favours male fitness will be bad when it occurs in a female, and vice versa. This means that the same gene experiences antagonistic selection, depending on which sex inherits it. Identification of the genes underlying sexual conflict and their place within the genome would probably advance our understanding of why some sexual conflicts can be resolved and some continue to escalate.

Another rapidly developing area in evolutionary biology is the study of epigenetics, changes to the genome of an organism within its lifetime that can be passed on to offspring at least for a generation or two. Scientists are increasingly recognizing that behaviours, fleeting though they might appear, can leave their signature in the genes. In many vertebrates, some sexually specific behaviours are programmed just before and immediately after birth, or during the transition from juvenile to adult, when the brain is exposed to different levels of hormones such as testosterone. The hormones also induce changes in the expression of genes related to sexually selected traits, which means that the environment can alter how the genes function. In turn, the gene alterations can be passed along to offspring. Such within-generation changes in genetic programming are also vulnerable to environmental insults such as pollution. Researchers are examining the ways that endocrine disruptors—chemicals that disturb the normal activities of hormones in the body—may eventually influence sexual behaviour and sexually related traits.

The very newest frontier in genetics is gene editing, a method for inserting new genetic material into an organism and removing material that is deemed undesirable. The techniques are usually discussed in the context of human health, so that a defective gene can be edited out and replaced with a functioning one, but the day is doubtless coming when such methods are used in experimental evolution, and hence could be relevant to sexual selection. Imagine the possibility of replacing a gene usually found in males with one found in females, and being able to observe the results in the organism as it develops. How such tools will be used to help us understand the evolution and genetic underpinnings of sexual behaviour is for the future to determine.

Finally, although we mainly try to understand sexual selection to gain greater knowledge of the world and its inhabitants, such knowledge can also be useful in our efforts to preserve biodiversity. Understanding the mating system and courtship behaviour of species in captivity can help us ensure that matings are successful. Sexual selection teaches us that mating is rarely random, and that simply putting together a male and female of the right species does not guarantee successful reproduction. In addition, as we noted in Chapter 7, allowing females to choose their mates may enable bad genes to be purged from a population, ultimately increasing the likelihood that a population or species will persist.

A great deal remains to be learned about the ways that animals engage over reproduction. Understanding how sexual behaviour evolves, along with the often-extravagant traits that accompany that behaviour, means understanding the complex ways that nature and nurture interact. Even Darwin was astonished by the power of sexual selection; as he said in *The Descent of Man* (emphasis ours), 'We shall further see, *and it could never have been anticipated*, that the power to charm the female has sometimes been more important than the power to conquer other males in battle.' Nearly 150 years later, we wonder what we might yet anticipate about the marvels of sexual selection.

Further reading

A number of academic books provide excellent detailed treatments of sexual selection:

Andersson, M. (1994). *Sexual Selection*. Princeton, Princeton University Press.

Arnqvist, G. and Rowe, L. (2005). *Sexual Conflict*. Princeton, Princeton University Press.

Eberhard, W. G. (1996). *Female Control: Sexual Selection by Cryptic Female Choice*. Princeton, Princeton University Press.

Rosenthal, G. G. (2017). *Mate Choice: The Evolution of Sexual Decision Making from Microbes to Humans*. Princeton, Princeton University Press.

Simmons, L. W. (2001). *Sperm Competition and its Evolutionary Consequences in the Insects*. Princeton, Princeton University Press.

The following are books written for a more general audience:

Birkhead, T. (2000). *Promiscuity*. London, Faber and Faber.

Cronin, H. (1991). *The Ant and the Peacock*. Cambridge, Cambridge University Press.

Ryan, M. J. (2018). *A Taste for the Beautiful: The Evolution of Attraction*. Princeton, Princeton University Press.

Zuk, M. (2002). *Sexual Selections: What We Can and Can't Learn About Sex from Animals*. Berkeley, University of California Press.

More specific recommendations, by chapter:

Chapter 1: Darwin's other big idea

Blaffer Hrdy, S. (1986). Empathy, polyandry, and the myth of the coy female. In R. Bleier (ed.), *Feminist Approaches to Science*. New York, Pergamon Press, pp. 119–46.

Milam, E. (2010). *Looking for a Few Good Males: Female Choice in Evolutionary Biology (Animals, History, Culture)*. Baltimore, Johns Hopkins University Press.

Dawson, G. (2007). *Darwin, Literature and Victorian Respectability* (Cambridge Studies in Nineteenth-Century Literature and Culture). Cambridge, Cambridge University Press.

Chapter 2: Mating systems, or who goes with whom, and for how long

Emlen, S. T. and Oring, L. W. (1977). Ecology, sexual selection, and the evolution of mating systems. *Science*, 197, 215–23.

Kvarnemo, C. and Simmons, L. W. (2013). Polyandry as a mediator of sexual selection before and after mating. *Philosophical Transactions of the Royal Society of London* B, 368, 20120042.

Thornhill, R. and Alcock, J. (1983). *The Evolution of Insect Mating Systems*. Cambridge, MA, Harvard University Press.

Shuker, D. M. and Simmons, L. W. (eds) (2014). *The Evolution of Insect Mating Systems*. Oxford, Oxford University Press.

Chapter 3: Choosing from the field of competitors

Kirkpatrick, M. and Ryan, M. J. (1991). The evolution of mating preferences and the paradox of the lek. *Nature*, 350, 33–8.

Mays, H. L. and Hill, G. E. (2004). Choosing mates: good genes versus genes that are a good fit. *Trends in Ecology and Evolution*, 19, 554–9.

Kokko, H., Jennions, M. D., and Brooks, R. (2006). Unifying and testing models of sexual selection. *Annual Reviews in Ecology Evolution and Systematics*, 37, 43–66.

Tomkins, J. L., Radwan J., Kotiaho, J. S., and Tregenza, T. (2004). Genic capture and resolving the lek paradox. *Trends in Ecology and Evolution*, 19, 323–8.

Chapter 4: Sex roles and stereotypes

Ah-King, M. and Ahnesjo, I. (2013). The 'sex role' concept: an overview and evaluation. *Evol. Biol.* 40, 461–70.

Janicke, T., Häderer, I. K., Lajeunesse, M. J., and Anthes, N. (2016). Darwinian sex roles confirmed across the animal kingdom. *Sci. Adv.* 2, e1500983.

Knight, J. (2002). Sexual stereotypes. *Nature*, 415, 254–6.

Chapter 5: Sexual selection after mating

Kvarnemo, C. and Simmons, L. W. (2013). Polyandry as a mediator of sexual selection before and after mating. *Philosophical Transactions of the Royal Society B: Biological Sciences* 368, 20120042.

Parker, G. A. and Pizzari, T. (2010). Sperm competition and ejaculate economics. *Biological Reviews*, 85, 897–934.

Parker, G. A. (2016). The evolution of expenditure on testes. *Journal of Zoology*, 298, 3–19.

Chapter 6: Sexual conflict

Rice, W. R. and Gavrilets, S. (eds) (2014). *The Genetics and Biology of Sexual Conflict*. New York, Cold Spring Harbor Laboratory Press.

Tregenza, T., Wedell, N., and Chapman, T. (2006). Introduction. Sexual conflict: a new paradigm? *Phil. Trans. R. Soc.* B 361, 229–34.

Wolfner, M. F. (2009). Battle and ballet: molecular interactions between the sexes in *Drosophila. J. Hered.* 100, 399–410.

Chapter 7: How sex makes species survive

West-Eberhard, M. J. (1983). Sexual selection, social competition, and speciation. *Quarterly Review of Biology*, 58, 155–83.

Marie Curie SPECIATION Network (2012). What do we need to know about speciation? *Trends in Ecology & Evolution*, 27, 27–39.

Panhuis, T. M., Butlin, R., Zuk, M., and Tregenza, T. (2001). Sexual selection and speciation. *Trends Ecol. Evol.* 16, 364–71.

Index

Index

SOCIAL MEDIA
Very Short Introduction

Join our community
www.oup.com/vsi

- Join us online at the official Very Short Introductions **Facebook** page.
- Access the thoughts and musings of our authors with our online **blog**.
- Sign up for our monthly **e-newsletter** to receive information on all new titles publishing that month.
- Browse the full range of Very Short Introductions online.
- Read **extracts** from the Introductions for free.
- If you are a teacher or lecturer you can order inspection copies quickly and simply via our website.